SCIENTIFIC-TECHNOLOGICAL REVOLUTION:
Social Aspects

with contributions by

Ralf Dahrendorf
Pyotr Fedoseyev
Ulf Himmelstrand
Radovan Richta
Yogendra Singh
Alain Touraine
Kazuko Tsurumi

and Introductory Remarks by
Ramkrishna Mukherjee

SAGE Studies in International Sociology   8
sponsored by the International Sociological Association/ISA

For information address

SAGE Publications Ltd
28 Banner Street
London EC1Y 8QE

SAGE Publications Inc
275 South Beverly Drive
Beverly Hills, California 90212

International Standard Book Number
0 8039 9961 5 Cloth
0 8039 9987 9 Paper

Library of Congress Catalog Card Number
76-22937

First Printing

Printed and Bound in Great Britain by
Biddles Ltd, Guildford, Surrey

CONTENTS

Notes on Contributors   6

Preface   9

1 Some Introductory Remarks   11
  Ramkrishna Mukherjee

2 The Scientific and Technological Revolution and
  the Prospects of Social Development   25
  Radovan Richta

3 Observations on Science and Technology in
  a Changing Socio-Economic Climate   73
  Ralf Dahrendorf

4 Social Significance of the Scientific and
  Technological Revolution   83
  Pyotr Fedoseyev

5 Science, Intellectuals and Politics   109
  Alain Touraine

6 Cultural and Social Contents of Scientific and
  Technological Revolution   131
  Yogendra Singh

7 Some Potential Contributions of Latecomers to
  Technological and Scientific Revolution:
  A Comparison between Japan and China   147
  Kazuko Tsurumi

8 On the Relevance of Historical Materialism in
  The Empirical Study of Advanced Capitalist
  Societies   175
  Ulf Himmelstrand

NOTES ON CONTRIBUTORS

Ralf Dahrendorf is Director of the London School of Economics, Director of the European Centre for Research and Documentation in Social Sciences, a member of the German PEN Centre, Honorary President of the Anglo-German Society and Senator of the Max-Planck-Gesellschaft. He has been a Professor of Sociology at three German universities and has been a Visiting Professor at several European and N. American universities. His publications include *Marx in Perspective* (1953), *Gesellschaft und Demokratie in Deutschland* (1965), *Essays in the Theory of Society* (1968), *Konflikt und Freiheit* (1972), and *The New Liberty: Survival and Justice in a Changing World* (British Broadcasting Corporation's 1975 Reith Lectures).

P.N. Fedoseyev has been Vice-President of the Academy of Sciences of the USSR since 1971. He is a foreign member of the Hungarian, Czechoslovak, Bulgarian, Polish Academy of Sciences and the German Democratic Republic Academy. He is also a Deputy of the Supreme Soviet USSR. From 1967-73 he was Director of the Marxist-Leninist Institute attached to the Central Committee of the Communist Party of the Soviet Union. His major publications include *Communism and Philosophy* (1962 and 1971), *Dialectics of the Present Epoch* (1966 and 1976), and *Marxism in the Twentieth Century* (1972).

Ulf Himmelstrand is Professor of Sociology at the University of Uppsala, Sweden. He was a fellow at the Center for Advanced Studies in the Behavioural Sciences at Stanford in 1968-69. Between 1964-67 he was Professor and Head of the Department of Sociology at the University of Ibadan in Nigeria. He is presently the Chairman of the Swedish Sociological Association and a Vice-President of the International Sociological Association (ISA).

Ramkrishna Mukherjee is a Distinguished Scientist at the Indian Statistical Institute, Calcutta. His previous appointments include being Professor of Indian Studies, Humboldt University, Berlin, Consultant to the Central Statistical Office, Ankara, and Chief Research Officer, H.M. Social Survey, London. His specialist interest is the historical and contemporary analysis of society. He is the author of about 100 papers, written or translated in English, German, French, Czech, Russian, Chinese, Japanese, and his mother-tongue, Bengali. He is also the author of twelve books, including *Social Indicators* (Macmillan), *The Rise and Fall of the East India Company* (Monthly Review Press), *The Dynamics of a Rural Society* and *The Problem of Uganda* (Akademie-Verlag, Berlin), *The Sociologist and Social Change in India Today* (Prentice-Hall), *Six Villages of Bengal* (Popular Prakashan, Bombay), *Data Inventory in Social Sciences: India* (Statistical Publishing House, Calcutta).

Radovan Richta is Corresponding Member of the Czechoslovak Academy of Sciences. He is particularly concerned with the social aspects of the scientific and technological revolution. He is the author or co-author of a number of studies, published in different languages, presenting an analysis of the processes involved in the scientific and technological revolution in different types of social systems and prospects opening up for man and the human civilization at present.

Yogendra Singh is Professor of Sociology, Chairman of the Centre for the Study of Social Systems, and Dean of the School of Social Sciences, at Jawaharlal Nehru University, New Delhi. He was previously Professor of Sociology and Dean, Faculty of Arts, Education and Social Sciences, University of Jodhpur, Rajasthan. He has taught at the Institute of Social Sciences, Agra, Rajasthan University,

Jaipur and McGill University, Montreal, Canada. He is the author of *Modernization in Indian Traditions,* and co-author of *Sociology of Non-Violence and Peace: Traditions of Non-Violence.* His co-edited works are *Sociology for India* and *Towards a Sociology of Culture in India.*

Alain Touraine is founder and Director of the Centre d'Etudes des Mouvements Sociaux (affiliated to the CNRS). From 1968-70 he was President of the Société Francaise de Sociologie, he is co-founder of the journal, *Sociologie du Travail,* and is a member of the council of the Max-Planck Institute. He is a Vice-President of the International Sociological Association (ISA). He is the author or editor of over twenty books and has contributed over 100 papers to various learned journals and academic institutes. His most recent publications include *Pour la Sociologie* (Seuil, 1974), *Lettres à une Etudiante* (Seuil, 1974), *Les Sociétés Dépendantes* (Belgium, 1976), *La Société Invisible* (Seuil, 1976) and *Au-delà de la Crise* (Seuil, 1976).

Kazuko Tsurumi is Professor of Sociology and Member of the Institute of International Relations for Advanced Studies on Peace and Development in Asia, Sophia University, Tokyo. Professor Tsurumi's publications include *Social Change and the Individual: Japan Before and After Defeat in World War II* (Princeton University Press, 1970), *Curiosity and the Japanese: The Theory of a Multi-layered Society* (in Japanese, Kōdan-sha, 1973), and, as a co-editor, *Adventure of Ideas: In Search of a New Paradigm of Social Change* (in Japanese, Chikuma Shobō, 1974).

PREFACE

The papers collected in this volume were presented in the first Plenary Session of the Eighth World Congress of Sociology, held in Toronto in August 1974. Professor Ramkrishna Mukherjee was the Chairman of the session, Professors Ulf Himmelstrand and Rudi Supek were Vice-Chairmen, and Professor Radovan Richta was the General Rapporteur.

Our original intention was to publish the papers under the joint editorship of Ramkrishna Mukherjee and Radovan Richta, but a number of difficulties emerged which have delayed publication and led to the withdrawal of Professor Richta as an editor, for reasons which are briefly indicated at the end of his paper (page 59 below) and also in the concluding sentence of Professor Fedoseyev's paper (page 106 below).

Professor Mukherjee was willing to continue as editor, but the Executive Committee of the ISA, after a very thorough discussion of the problems at its annual meeting in May 1977, decided that it would be more appropriate to publish the volume as a collection of individual papers, in order to emphasize as strongly as possible that each author is responsible *only* for the views expressed in his own paper. The volume as it now appears, therefore, contains the papers contributed to the Plenary Session (including the opening address delivered by Professor Mukherjee in his capacity as chairman of the session) with revisions by their authors; and in addition a revised and shortened version of the comments which Professor Himmelstrand made on the general theme of the session.

We regret very much that publication of this volume has been so long delayed, and that it has not been possible to present it in the form which was first envisaged. But we were convinced all along, as were the contributors themselves, that it was extremely important to publish these papers in one form or another, and we are glad that a solution to the difficulties has now been found. The Plenary Session to which the papers were contributed had an exceptional significance inasmuch as it was intended to set out the general theme of the Congress and to provide an analysis of some fundamental questions concerning the interrelations between scientific and technological advance and the development of present day societies. We hope that the reader will find here, amid the considerable diversity of approaches, elements of knowledge and ideas which will be of use in the continuing effort of sociological investigation to throw light upon these vast and urgent problems.

Tom Bottomore, *President*
for the Executive Committee, ISA
May 1977

1

SOME INTRODUCTORY REMARKS

Ramkrishna Mukherjee
Indian Statistical Institute

I

The world of today appears to present two sharply contrasting pictures: one of frustrating pessimism, the other of buoyant optimism. Never before has crisis been so acutely manifest in the life of the people in virtually all parts of the globe. Its expression may be different in different societies: economic and political in the 'developing' societies; ideological in the most ideologically structured 'developed' societies; or social and cultural in the most economically prosperous 'developed' societies. In respect of the physical environment and ecology, the crisis has also attained a world dimension. Apart from the current energy crisis, which has international repercussions, it has led to a questioning of the 'limits to growth' irrespective of inter-societal variations.

Thus, in whatever form and whatever place it may erupt, a world crisis is now manifest. At the same time, never before in the history of mankind has so much rapid and outstanding progress been made in the fields of science and technology. Although the progress is most manifest at the two epicentres of world power and prosperity it has spread over the whole globe. The world has become small, and the people have the conviction that man is now in a position to subjugate Nature instead of merely adapting himself to its prescriptions. Generally, this is regarded as the revolutionary portent of science and technology in the present era.

Now, what are the links between these two contrasting pictures of the contemporary world? The crisis can be solved by the application of knowledge to optimize the relation between ends and means; and knowledge is systematized in science, applied through technology. The contrast, therefore, raises several *operational* questions of immediate relevance:

(1)   Will the crisis be resolved by the scientific and technological revolution? Alternatively, if the *social*   and *national* revolutions are the means to resolve the crisis in different societies, what is the relevance of the scientific and technological revolution in that context?

(2)   Which forms of the crisis are connected with which aspects of the scientific and technological revolution in different societies? Correspondingly, does the apprehension of 'limits to growth' imply a crisis in science and technology per se?

(3)   In an analogous context, has the ideological crisis, or the crisis in culture, any relation with the development of science and technology?

(4)   Lastly, following from a positive but critical appraisal of the scientific and technological revolution in the world society, what are the *soft spots* in the respective societies that make it possible to promote the development of science and technology and what are the *hard spots* that retard the process?

In the light of these questions, the contrast between crisis and achievement also raises problems of *explanation:*

(1) Is the crisis the resultant of conflicting social systems? In that case, can scientific and technological revolu-

tion supersede the crisis and thus override the basic contradictions between the two major world systems: socialist and capitalist? Also, will the process resolve, concurrently, the dissensions within the socialist and the capitalist spheres?

(2) Do the 'limits to growth' present an inevitable phenomenon in the contemporary and future world perspective? Alternatively, can the revolutionary fervour of science and technology demolish this spectre haunting a distinctive sector of the world intelligentsia?

(3) In which specific contexts, therefore, should we speak of scientific and technological 'revolution' — and not merely of 'development'?

In addition to its immediate relevance and explanatory significance, the contrast involves other questions of fundamental importance:

(1) Is the crisis the expression of contradictory human values?

(2) Can the scientific and technological revolution generate knowledge in such a manner that it will solve the controversy over the material or the spiritual basis of man's existence?

(3) Conversely, will science and technology lead more and more to the *alienation* of man from society and nature?

(4) In sum, what is the purpose or end of the scientific and technological revolution?

In the ultimate sense, therefore, the contrast leads us to a clarification of the concept of 'social development'. Is it from 'matter' to 'mind' according to the Marxist schema, or from 'mind' to 'matter' according to the ideological thinking of Hobhouse, Ginsberg, Mumford, and others? Alternatively or additionally, does it rest primarily upon the fact of 'social existence' as formulated by such thinkers as Morgan, Durkheim, Simmel and Spencer, or upon the assumption of 'social consciousness' as propounded by

Rousseau, Pareto, Tönnies, Weber and others. These questions are basic to the contemplation of the issues posed above because the world of tomorrow will be shaped by the answer to two questions: 'What *will* it be?' and 'What *should* it be?'.

II

Our frame of reference for considering this series of questions is the *social aspects* of the scientific and technological revolution. The elements constituting the frame are social structures and processes, systems and values, actions and their management. We should discuss with reference to this frame why we speak of 'revolution' and not of 'growth' or 'progress'; that is, how exactly the revolutionary force of science and technology can conquer nature and surpass the apprehended 'limits to growth' in a world perspective. Furthermore, we should clarify the distinctions among the similar phenomena of 'social' revolution, 'national' revolution, and the 'scientific and technological' revolution, and at the same time establish a relation among them.

Should we regard social revolution as being essentially an *intra*-societal phenomenon, and national revolution as occurring at the level of a total society, while the scientific and technological revolution is to be apprehended in an international perspective? In that case, what are the possibilities and the procedures for *internalizing* the revolutionary course of science and technology as it is presently considered under such labels as 'ethno-science' and 'ethno-technology'? We should examine in a similar context the phenomenon of 'unequal exchange' of both material and non-material commodities within and between societies. For this phenomenon is seen today to hinder social development, however we may formulate the concept. It is not only *germane* to any crisis situation but also refers directly to the development of science and technology on a world-scale.

Obviously, dealing as we are with a multitude of facts and phenomena, our discussion of the scientific and technological revolution is likely to go beyond any one presentation of the subject-matter. It may incorporate other elements besides those mentioned above. Moreover, a particular presentation may not consider the crisis situation at all, but may dwell exclusively upon the scientific and technological revolution per se. In all such eventualities, however, the range of discussion needs to be organized under a logical construct. One such contruct would comprise three schematic distinctions: the *social function*, the *social basis*, and the *social content* of the scientific and technological revolution.

The social function of science and technology can be appraised for mankind as a whole, irrespective of the inter-societal and historical variations. Historically, of course, technology precedes science. Man made fire before he systematized his knowledge about the phenomena of heat and light. But neither science nor technology develops without interaction; and by their inherent characteristics both are of *universal validity*. Therefore, in the context of the world at large, the contradictions and the consequences of scientific and technological revolution become a distinctive focus for examination — such as the use of atomic energy for war or peace, and the application of science and technology for space research or for the betterment of mankind.

Now, while the role of science and technology may be appraised for mankind as a whole, their *relevance* at a particular point in time and/or space does not always synchronize. In the time-dimension, technological development has not invariably led to scientific development, or vice versa. One such example is the distinction found in an overall perspective between Europe, on the one hand, and India and China, on the other, in the pre-eighteenth-century period. In the space perspective, also, an unequal development of science and technology is currently observable among the nation-states, both developing or developed. Moreover, in this context, the inter-societal exchange of scientific and technological knowledge may be either for mutual benefit or asymmetical. The social basis of scientific and technological revolution, therefore, becomes another distinct focus for examination.

Lastly, the use and applicability of science and technology would be specific to a 'people'; that is, to an *object* in the sequence of *place* and *time:* the three dimensions of variation which have been stated in Indian philosophy in the phrase *sthānā-kālā-pātra,* in order to depict reality precisely and comprehensively. The people constitute a dimension of variation to appraise the *specific efficiency* of the scientific and technological revolution because it can be promoted or retarded by the intra-societal ethos, structure and processes, and the corresponding actions. Therefore, the social content of scientific and technological revolution becomes the third distinctive focus for examination.

These three foci help us to appraise the social aspects of the scientific and technological revolution in terms of its universal validity, particular relevance in the place-time dimensions, and specific efficiency in the context of a people. But they should not be treated as mutually exclusive. Relations must be established between the social function, the social basis, and the social content of science and technology in order to show the significance of the two phenomena acquiring the character of a 'revolution' for present-day society.

This should also provide us with the knowledge to resolve the contrasting pictures of crisis and achievement in the world today. The social, cultural, economic and/or ideological discriminations *within* a society are likely to lead to social revolution. In respect of these characteristics, the unequal exchange *between* societies promotes national revolution. And these two revolutionary accomplishments are necessary conditions for the success of science and technology in a society, as we can see with particular reference to the Renaissance in Europe and the remarkable achievements of the US and the USSR in recent times. It is also apparent that the internalization of science and technology, in relationship with the ethos and structure of a society, can stimulate, and even ensure, social and national revolutions. There are several relevant examples in present-day Asia, China, Japan, India and, notably, Vietnam. Thus, where social and national revolutions are not yet fully accomplished, and further rapid changes are on the agenda, the scientific and technological revolution may provide the impetus and also lead to their definitive achievement, provided it is directed in tune with the society and the people concerned. Thus, the *social aspects* of the scientific and technological revolution may synchronize with social and national revolutions and usher in an era of everlasting peace, continual progress and consistent prosperity on a world-scale.

III

From this perspective, the role of values in science and technology attains a crucial importance: whether science and technology are at all desirable for the development of human society; whether some of their characteristics are good for the progress of humanity, while others are bad; and to what extent the 'good' characteristics should be pursued to harmonize the social development of mankind with nature.

As I noted at the outset, science consolidates knowledge, and technology applies the body of knowledge for one purpose or another. Obviously, this purpose, whatever it is, cannot be value-free: we must regard it as 'desirable' or 'undesirable', 'good' or 'bad'. That is, we assign distinct value significance to scientific and technological development with reference to their diverse aspects considered as good or bad. The role of value in science and technology is thus multi-faceted.

Now, who decides what is desirable and good, and what is undesirable and bad? In the last analysis, the decision rests with each individual; but it is formulated and executed at 'group levels'. The smallest of these levels is a family, a cultural or a political group, or any such societial unit. The larger level is that of a society, a country, a nation-state. And the omnibus level is that of mankind as a whole. But these levels are not hierarchically structured in one dimension with reference to the role of value in science and technology. What any one group at any one level considers as good, another may regard as bad. The role of value in science and technology is, therefore, not only multi-faceted but also multi-dimensional.

However, from two diametrically opposed viewpoints any consideration of value in science and technology appears to be redundant. One of them is that man, in order to be happy, should eschew science and technology, and be in communion with Nature. The other is that man, in order to be happy, should have the means to release all his potential energy which requires the utmost development of science and technology. Thus the first viewpoint seems to negate science and technology, and so also their relevance to values, while the second viewpoint simply equates them with what is valuable. But the appearance is deceptive or inconsequential.

Even when communing with Nature man must produce in order to live, protect himself from the environment, and propagate the species. So he cannot do without a systematic accumulation of knowledge and its application through technology, however 'simple' and 'unsophisticated' the scope of science and technology may appear to be. Thus, even the first extreme viewpoint must evaluate science and technology in relation to what is desirable or undesirable. The second viewpoint, of course, does not intend either to reject Nature or to make man into a robot. Yet this anxiety lurks behind the appreciation of science and technology for the progress of mankind. As a result, we have different views on social development from, say, Marx to Mumford; and these views are popularly placed under polar-opposite labels of the 'material' or the 'spiritual' basis of human progress. At the present time, attempts are made to *standardize* the role of value in human progress, of which scientific and technological development is almost invariably reckoned to be an essential component.

International organizations such as the United Nations and UNESCO tend to arrive at a general consensus on the minimally desired goals which automatically require scientific and technological development. These goals are that men should enjoy good health, adequate nutrition, education, shelter and so on. But even when a general consensus is attained among the members of the United Nations on such goals, the perennial question remains: What happens *after* attaining the minimal targets? And the question remains relevant as to *how* to attain these goals. Values in this context may be so different that one person may believe that 'capitalism springs eternal in the human breast' while another may consider that 'proletarians of the world should unite'. Obviously, the role of value in scientific and technological development cannot be standardized in this manner, especially because development should be regarded as a never-ending process in human existence.

On the other hand, any value significance imposed upon scientific and technological development would be impractical and even injurious to the people concerned. The concept of 'modernization', fashionable these days, represents one such valuation. Its proponents clearly regard the dominant value-system of Western Europe and North America as playing the deciding role in the concept of 'modernization', in which the principal component is the characteristics of scientific and technological development pursued and achieved in this Atlantic region. But the imitative and non-contextual application of the concept of 'modernization' and its particular 'blue-print' of scientific and technological development have not benefited the 'developing' societies.

In most societies of the Third World the 'modernization ideals' have led to the prosperity of the imported and indigenous elites at the cost of misery for the masses. The achievements of the First United Nations Development Decade, which was geared to these ideals, have not only not spread to the masses, but the spectre of an acute crisis in these societies now haunts the world. Accordingly, the Second United Nations Development Decade has announced a programme for the eradication of poverty and inequality in the 'developing' societies. Moreover, influential and distinctively articulate sectors of the people in these societies do not subscribe to the encapsulated 'modernization ideals' to bring about the social and/or national revolutions.

Thus, neither a minimally agreed nor an enforced value significance of scientific and technological development will serve humanity. The role of value in this context must be specific to the people and the society concerned. Hence, in the present state of our knowledge in theory and action, we cannot speak of *one*, or *the only* role of value in scientific and technological development. It has to be a matter of *diagnosis* in a place-time-object bound field in which there is diversity of possible value significance.

IV

This diagnosis, however, is not to be applied to a random situation. In fact, if we did so by following the path of gross empiricism, our understanding of social reality with reference to science and technology would remain just as limited as if we have imposed a theory as dogmatists and doctrinaires. We may, then, describe or explain a situation from a particular orientation, but our answers to the two questions, 'What will it be?' and 'What should it be?' will not be precise, unequivocal, or comprehensive. For such answers would require a probabilistic evaluation of all available and possible explanatory propositions in which the role of value in science and technology is involved.

On the other hand, if we examine the role of value in science and technology in its substantive details, we do not encounter a random situation. The views on social development may appear to be polar-opposites in extreme situations, but the differences among them are matters of emphasis and not the negation of either the material or the spiritual aspects of the peace, progress and prosperity of humanity. At one end of the scale the material basis of society denotes a particular form of social existence which, in its turn, evolves a specific kind of social consciousness and upholds the spiritual basis of society. At the other end of the scale, the spiritual basis of society denotes a particular form of social consciousness which, in its turn, evolves a specific kind of social existence and upholds the material basis of society.

In neither case is the need for scientific and technological development ignored. Only the role of value in each context assumes a distinctly different character, of which an extreme manifestation is recorded by either one of the viewpoints we have stated.

Since we can thus ascertain the respective bases of emergence of different valuations, we need not consider the multi-dimensionality of value in scientific and technological development to be indiscriminate or indeterminate: the valuations can be systematized into a constellation. Furthermore, the forms of expression and execution of the different valuations are not incoherent. They can be identified, with reference to the nature and extent of differences in the valuations, while appraising the dynamics of a society and, eventually the world society. For no society in the world is insular and self-contained.

What we need, accordingly, is to examine systematically the distinctions and interrelations among the differential valuations at any level of comprehension and execution of science and technology in the place-time-object dimensions. Concurrently, we should ascertain their actual influence and effect on a society, and the world society, from an empirical appreciation of the social dynamism. Such an interaction between 'theory' and 'empiricism' should reveal unequivocally the relevance and the *relative* efficiency of particular valuations to depict the social reality. Hence, the process should prove the inconsequential character of some value standpoints, and the decisive efficiency of a particular valuation or of a specific set of valuations in terms of their sequential cause-effect relations in the set.

In this way, by means of a systematic and unambiguous process of rejection and consolidation of valuations according to their actual relevance and efficiency, the role of value in science and technology may be eventually standardized. At any rate, the proposed process of systematization and empirical appreciation of the role of value should lead us to an objective assessment of the necessity and efficiency of the scientific and technological revolution for one society, at one end of the range, and for the world society, at the other.

Thus, we may proceed from the bottom (as it were) by taking note of the 'singularity' of each society, next the 'particularity' of a number of similar societies, and finally attain the summit of 'universality' in due course. In contrast with this procedure, any ad hoc or dogmatic formulation of 'universally' valid, relevant and necessary values for science and technology would be simplistic or an imposition on our objective appraisal of the social reality. And, since the 'one world' concept is intrinsic to the revolutionary potential of science and technology in the present era, sociologists may consider it their specific duty to mankind to explore the social aspects of the scientific and technological revolution in the manner briefly proposed here.

2

THE SCIENTIFIC AND TECHNOLOGICAL REVOLUTION AND THE PROSPECTS OF SOCIAL DEVELOPMENT

Radovan Richta
Czechoslovak Academy of Sciences

There are many signs that the 1970s will put sociology, its cognitive powers and practical value, to a most stringent test. The development of society, marked by sweeping revolutionary transformations affecting its entire system, places considerable demands on science. The ever-expanding, conflict-ridden scope of possibilities opened up by the advance of the scientific and technological revolution in different social systems has confronted sociology with an urgent task — to clarify the basis, content and social function of the process before changes in the material conditions of man's life transgress a certain critical point.

As to intensity, the growth of science no doubt outstrips the development of many other areas of social life. The gradual introduction of automated production processes affect the traditional mode of man's participation in the process of production as well as the content of work and the structure of occupations. The formation of a new power and material basis interferes with the established dependence of human life on natural resources. Cybernetics facilitate the creation of automatic control systems. Man is confronted with an ever-mounting flood of events and information. Science and technology have invaded family life as well as emotional relations. They have contributed significantly towards a substantial extension of man's life. The ever-growing number of means affecting the environment and the interaction between society and Nature are now doing away with the belief in the immutability of climate. The emerging 'biological technology' and 'biological engineering' have opened up prospects for the artificial reproduction of live tissues and organisms as well as the possibility to interfere with the genetic code of all living organisms and with man's mental make-up. The astounding potential inherent in the projects envisaging the utilization of space and oceans foreshadows new chances for the transformation of a number of processes specific to terrestial life. The means of mass destruction are transgressing the boundaries of absurdity thus making peaceful coexistence absolutely indispensable.

It would be difficult today to find an area of social life that remains neutral towards and unaffected by the scope and frequency of the sweeping changes currently underway in the fields of natural science and technology. At the same time, the wide-ranging and most significant social consequences ensuing from the application of science today relate to the tasks of a science-based control of social development, a science-based revolutionary reconstruction of society.

The problems posed by the scientific and technological revolution are not only a matter of concern for specialized research; they challenge the very foundations of social theory, the general concepts of sociological inquiry. Once again at least three questions of this type will have to be answered:

(1) What exactly do contemporary science and tech-

nology represent as phenomena of social life?

(2) Where do trends of contemporary scientific and technological progress come from, and in what way can its prospects be controlled and geared towards the attainment of social goals?

(3) Which type of society is capable of mastering the scientific and technological revolution? What are the social relations that provide an adequate basis for the utilization of such achievements as a universal computer network, or again, interference with the genetic code of living organisms, the control of climate and the artificial reproduction of the give-and-take between society and nature?

THE SCIENTIFIC AND TECHNOLOGICAL REVOLUTION AND ITS SOCIAL CONTENT

One of the most exacting tasks of social analysis is the identification of the nature of a process which does not exist in a 'pure' laboratory form but always within the context of different social systems differing in the stage of development attained and — most important of all — displaying an antagonistic social nature. In the process of the scientific and technological revolution, these circumstances operating under the surface of a variety of overlapping factors undoubtedly create extremely difficult objective conditions for sociological research into the process of the scientific and technological revolution.

The history of social theories provides only few examples of research comparable in scope and frequency with inquiries into the social aspects of the scientific and technological revolution in the early 1970s. Within the period between the Seventh and Eighth World Congresses of Sociology alone, several hundreds of publications on this topic have appeared throughout the world, while studies on the subject run well into the thousands.

In profile and results, the research displays fundamental differences. Illustrative in this respect are the results of sociological research in those Western countries that, due to specific historical circumstances, were the first to face the challenge of the problems attendant upon the present-day dimension of the development of science, technology, and productive forces; an example being provided by the United States which displayed the highest productivity of labour and spent, up to the end of the 1960s, more than half of the total world expenditure on research and development.[1] With the development of science and technology, the relevant literature often reflects the rapidly alternating concurrent or successive periods of fascination and revolt, 'heroic indifference' and frustration of research concerned with these realities.[2] The 1960s, no doubt, widened the scope of inquiry into the growth of science and scientific information conducted along lines analogous to the study of 'gas in thermodynamics' (Price). Inner changes in the theoretical conceptions of science came within the purview of research (Kuhn). Individual revolutionary processes in technology ('Computer revolution', 'Cybernation revolution', 'Energy revolution', 'Biological revolution', etc.) began to attract attention.

This, however, did not result in a coherent theory encompassing all these changes, presenting them as a complex process and defining their role in society and its development. In spite of all this, to quote Merton, the fundamental issues of 'the modes of interplay between society, culture and science are with us still.'[3] The general feeling of gradual loss of control over the contemporary development of science and technology in its implications for the life of man and society, and the ensuing apprehensions about transforming the world, mark in one way or other the vast majority of studies on scientific and technological revolution in the capitalist countries. The term 'scientific and technological revolution' itself did not acquire a more precise conceptual explication and gave rise to occasional doubts and speculations.[4] Nevertheless, each new advance confirms that at present this term can hardly be dispensed with.[5] In its most frequent usage in the Western sociological literature, the term 'scientific and technological revolution' refers either to changes[6] arising as a result of major scientific discoveries, or again to a set of social, economic and technological changes,[7] thus making the term identical with earlier concepts of the 'second' or 'third' industrial revolution. In this sense major changes within the existing capitalist system are then postulated.

Difficulties connected with the evaluation of contemporary scientific and technological change have their own methodological aspect. How could a truly revolutionary innovation be identified? It is not the technological parameters alone that offer the answer. The solution is implicit in the social functions specific to a given type of technology and, above all, in its basic function in the reproduction of human life. An adequate framework for the solution of this problem is provided by the Marxist theory of the dialectics of *productive forces* and productive relations defining both the inner composition and the social role of productive forces. In these terms, it is precisely the type of scientific and technological progress that changes substantially the nature of the productive forces of society, their structure and dynamics that is to be specified as revolutionary. It was the analysis of changes in the social role of science that motivated Bernal's proposal to base the sociological analysis of changes connected with the present-day ventures of science and technology on the concept of 'scientific and technolgical revolution'.[8] It is precisely in connection with

the role of science in the development of productive forces
of the new society that right since the mid-1950s Soviet
analyses have employed this concept.

This line of thought proved to be productive. The concept
of scientific and technological revolution has gradually
become an 'indispensable and permanent feature' of analyses
concerned with the present-day social facts and realities of
the socialist countries.[9] As is evident from the relevant
literature of the early 1970s, Marxism-Leninism arrived at
a coherent conception of the main characteristics of the
scientific and technological revolution. Scientific analysis
based on these principles makes it clear that not only the
growth of science is at issue but, above all, the process of
its transformation. Once a closed, exclusive area on the
margin of society, science becomes an immense organism
operating in all spheres of social life. A *new type of science*
is gradually emerging, differing[10] considerably as to social
function and theoretical and methodological principles
from both ancient knowledge (focusing not on production
but on the cultivation of the virtues of the free classes)
and the New-Age science originating during the Renaissance
and the Industrial Revolution, marked by the utilitarian
drive towards extraneous interest. The logic underlying
this development is a tendency towards the integration of the
process of cognition (and transformation) of the world with
the process of self-cognition (and transformation) of the
society — the conception of science evolved by Marx, Engels
and Lenin.

Similar *fundamental changes* are in progress in the field
of *technology.* Contrary to the essentially machine (man-
operated) type of technology that until recently formed the
core of industrial (factory) production a new technological
basis[11] relying on automation and cybernetics is gradually
coming into being. This sets in motion the entire machinery

system — thus shifting the focus of man's nature-orientated activities from the mere transformation of nature to the control over the entire process of transformation of nature by society.

Finally, *interrelations* between science, technology and production are set in motion. Production, of course, continues to enforce the development of technology, and technological requirements stimulate the development of science. What is new here is that, at a certain stage of the development of material condition, scientific progress functioning as a force mediating the development of technology and production creates a new, ample scope for feedback phenomena, thus gradually anticipating technology and production — a feature typical of the process of scientific and technological revolution.[12]

Crucial for the specification of the social nature of the scientific and technological revolution has been the identification of the inner link between present-day development of science and technology and those changes in *productive forces* described by Marx and Lenin as the general application of science in the form of an 'immediate productive force', as the transformation of production 'from a simple working process into a scientific process',[13] as the gradual transition of specific functions exercised by man in the process of production to the functions of technology.[14] This provided an adequate basis for the definition of the scientific and technological revolution as a complex of fundamental changes affecting science and technology, their system, interrelation and social function, leading to a universal change in the structure and dynamics of the productive forces of society, altering man's role in the system of productive forces on the basis of a complex application of science as a social productive force pervading all other components of productive forces and remoulding the entire technological basis of human life.[15]

An identification of the specific features of the scientific and technological revolution provides an insight into the interrelations between present-day advances of science and technology and the development of the socio-economic system. The main sphere of operation of science and technology within society is mediated through changes in the inner composition of the society's productive forces — primarily through changes in their social and technological components[16] quantitatively reflected in the growth of labour productivity. If they ignore the mediating link of the dialectics of productive forces and production relations, sociological inquiries into the relation between the scientific and technological revolution and social progress are bound to oscillate between conceptions of mutual independence of social and technological development, and mere reduction of one factor to the other.

The transition from productive forces based on the opposition: 'machines thought-up by human intellect' versus 'simple manual work'[17] to a complex pervaded in all its components — whether technological or human — by science, brings about 'fundamental qualitative changes in the technological basis of production and its management, as well as changes affecting man's role and function in the process of work'.[18] The process releasing labour force from the mere operation of machines, for control and preparatory activities entails an *organic linkage of science and production* — the main dimension of the development of productive forces under the conditions of the scientific and technological revolution. This implies the application of science in the technology of production and the corresponding utilization of organizational and professional standards (this could be regarded as the 'narrow range' of development) and, on the other hand, the appropriation of science by the aggregate working people — both through the application of science in technology, organization and qualification — and at the same time through the development of the creative potential of all working people (the 'broader range' of development).[19] Under conditions of the scientific and technological revolution each of these two cycles within the development of the productive forces has its own specific significance and role. While the former is no doubt typical in the first stages of the development of productive forces (a limited form of application of the scientific and technological revolution), the significance of the latter cycle increases in the subsequent stages of the process. A fully fledged realization of the scientific and technological revolution is dependent on an adequate development of the latter sphere.

Productive forces do not, of course, exist outside the set of social relations specific to a given social system. The interaction of the components within the general structure of productive forces is mediated through the operation of production relations representing the social forms of motion of the productive forces and determining the basic social agent of their development. In this way the dynamic potential inherent in the given structure of productive forces is actualized in two different ways corresponding to two different modes of the social functioning of science:

(1) The *narrow range* corresponds to a pattern of application of science as evolved in capitalist society (in the form presented, for example, by Max Weber) with science representing a sub-system separated from labour and opposed to labour. Thus the use of science is primarily a matter of its technological application and the corresponding organization in the form of 'capital fixe'[20] imposing control over the working masses (productive forces of the social individual). Under this system each application of science is based on the activity of a special elite of the 'elect' opposed to the masses

of the 'non-elect' who are not the bearers in the process of rationalization but its mere objects. The development and creative activity of the working people are not considered as constitutive elements of this type of progress of human civilization.[21] Science of this kind — as openly admitted by Weber — is indicative of class contradictions and reproduces them on an ever-widening scale with each new advance. 'Eternal' and 'natural' as it may appear this form of social utilization of the scientific and technological revolution is a historically limited one.

(2) The *broader range,* opened up by the socialization of the means of production, the elimination of capital and the barriers of private ownership between different components of the productive forces, allows for the complete linkage of science as 'universal labour' (Marx) with the total labour of all individuals.[22] The application of science in technology involves no more than a section of the process within which the workers' collective as an active social agent appropriates science, transforming it into a link mediating the all-round development of the powers, potentialities and activities of the entire working people and, in the final analysis, linking science in all its applications as 'general social labour' with the labour of all individuals.[22] Under socialism this in turn becomes the primary mode of development in the sphere of productive forces and the creation of 'capital fixe' — '. . . this capital fixe being man himself'.[23]

It is evident that the early stages of the scientific and technological revolution can occur in different social systems displaying an adequate material base. But its full implementation is only possible within the broad range, while the petrification of the narrow range necessarily results in the deformation of the entire process and a premature exhaustion of the developmental potential.

CHANGES IN THE FUNCTIONS
OF SCIENCE IN SOCIETY

The advent of the scientific and technological revolution confronted sociology with new aspects of the problem of the development of science and its relation to society. Numerous complaints concerning the 'continued neglect'[2 4] of this area of research, raised by sociologists in the West, are indicative of the profound cause of difficulties attendant upon this issue. It emerges that the real form of the development of science as a system 'insulated'[2 5] from all other forms of social labour and essentially designed to improve technological facilities controlling workers' activities in the production process is reflected in the theoretical approach considering this markedly historical, temporary mode of existence and the autonomous development of science as a natural and eternal state of affairs.

It was precisely the real subordination to the interests of capital that gave the traditional system of science, originating in the period of the Renaissance and Industrial Revolutions, the character of a mere instrument, designed to achieve extraneous utility — the character of 'value-free rationality' separated from its social prerequisites, eliminating the social subject from social theory, mediating technological manipulation of things and of man as a thing,[2 6] pervading man's life in the form of a 'disenchanting' calculus geared towards domination, unable — to recall Tolstoy's criticism — to provide an insight into the sense and aims of the life of man and society. This conception of 'rationality' which has not only provided the basis for research into the social impact of science but, to a certain extent, for the theory of social development in general, actually offers a theoretical transposition of the 'narrow range' of application specific to science within the capitalist system. The inverse aspect of the 'insulation of science' in capitalist countries can be traced in the difficulties besetting some theories of social development.[27] Under the conditions of the scientific and technological revolution, the constraints thus imposed on science and the role it plays in the life of society gradually acquire the character of a restrictive framework, completely lacking a dimension conducive to achieving the decisive point in the development of productive forces, i.e. the linkage of science as a productive force with the all-round development of the faculties, abilities and activities of the social individual.

The assumption that present-day science (and scientific and technological progress) is the exclusive concern of a scientific and technological elite — one that may have seemed relevant decades ago — stems in fact from a professional illusion; under the conditions of the scientific and technological revolution, this assumption proved just as untenable as the fallacy — once destroyed by the Industrial Revolution — that national economy is the business of court economists, or again, the claim — dispelled during the Renaissance period — that emotional life is simply the concern of poets. Present-day science represents 'general intellect' (Marx), social cognition and a social productive force. Each advance of science is the result of the entire development of preceding generations and the co-operation of contemporaries. The development of science indicates the degree of mastering society's general productive force, the level of development of the intrinsic potentialities of the social individual.

Seen in this light there is a striking disproportion in sociological literature focusing on the impact of scientific and technological progress on society on science and technology.[28] This fact, too, is evidently closely connected with the specific social reality of science and the specific type of development of productive forces under capitalism. Within this framework, the onset of the scientific and technological revolution stirs the surface of life as a spontaneous, uncontrolled and uncontrollable current, as a 'continuous innovation and expansion of rationality'.[29] The specific expression of a narrowly orientated and conflict-ridden development of productive forces under capitalism frequently generates conceptions of social development which describe scientific and technological progress in terms of a demoniac 'megamachine' 'feeding on itself', of a 'fatal process', 'independent of man and society', 'infinite in space and time', governed by mere intrinsic 'principles of efficiency', and defying any attempt aimed at interfering with its course. Contrary to the 1950s and early 1960s, in the sociological literature of the past decade this aspect acquired a dramatic dimension as reflected in the bizarre analogies of the type of 'an engine tearing along with no engine-driver or brakes', or 'the ride on a tiger'. The belief that science, technology and economic growth will solve the fundamental issues of life gradually gave way to 'a feeling of uneasiness vis-à-vis the prospects it opens up'.[30] The only parallel offered by sociological literature with regard to the frustration and disillusionment creeping into some of the present-day analyses of the social consequences of scientific and technological progress, particularly in connection with the ecological crisis and the hot debates surrounding the 'quality of life', is the reaction to the crisis phenomena of the 1930s.

All these facts engender the tendency to describe technology 'permitted to follow its own logic . . .' as 'a cancer-like growth, eventually endangering the structured system of individual and social life',[31] as 'the ultimate author of these ills' and especially 'too much technology too fast'.[32] In one way or other, scientific and technological progress 'got beyond' society's control — this conclusion not only underlies a number of romantic protests against involving man in the 'intrigues' of his ever-increasing material power[33] but is also implicit in a number of anti-technical projects of 'Counter-Culture',[34] negatively indicative of the blatant absence of the human aspect in specific types of techno- logical progress in capitalist countries; it is likewise present in the Stoical appeal of some bourgeois prognoses adhering to the principle that technological change sets in fatally whenever feasible;[35] a similar conclusion underlies the no less negligible section of the 'technology assessment' trend,[36] still lacking a solid sociological background. Finally, all this has a profound impact on bourgeois social theory, reversing some of its traditional starting points[37] and raising the question whether 'society is to be the servant or the master of the instrument it creates'.[38]

What is behind the fact that increasing knowledge, tech- nological advance and the ensuing rise in labour productivity — so frequently described as an incarnation of 'rationality' — clash with the vital issues of man's and society's life?

Understandably, the reassessment of the implications of scientific and technological progress has to begin with a critical analysis of the social realities accounting for the contradictory nature of 'rationality' entailed in the fact that while science and technology 'become more autonomous and more powerful, men experience themselves as less potent, less in control of their own destinies'.[39] Techno- logists offer evidence that, at a relatively advanced level of labour productivity in developed capitalist countries, the achievements of the scientific and technological revolution provide adequate scope not only for minimizing the pro- portion of monotonous, exhausting work procedures but also for a systematic, gradual transformation of the content of basic production activities for millions of working people.[40] This potential, however, is largely neglected and remains untapped by 'rationality' specific to these social conditions — thus perpetuating a situation in which the working masses still find their work a dreary burden.[41] Sociological analyses, however, have traced similar disparities in a number of other areas of contact between social and technological change.[42] C. Wright Mills and D. Riesman were among the first scholars in American sociology who raised the question — obviously related to the onset of the scientific and technological revolution — whether the existing scientific and technological progress in the US is really compatible with rationality, and whether the 'conspicuous production of our corporation' and the 'conspicuous science and technology'[43] are the only possible alternatives for the progress of civilization at this advanced stage of the development of productive forces. This criticism introduced the idea of changing the *deformed course* of development in science and technology, to reverse the orientation of 'technical impetus' and 'redesign' its channels.

Nevertheless, the measures taken in highly developed capitalist countries to bring the social after-effects of the scientific and technological revolution under the control of 'rationality' led, at each new stage, to the re-emergence of the same contradictions; thus the gap between the rapidly expanding technological facilities and the stagnant social and human conditions is contantly widening. Due to the fact that with each new advance of the scientific and technological revolution it is precisely the social factors that gain in significance, further progress along these lines is envisaged even by optimists as a 'planned muddling through'[44] these increasingly complicated conditions. This is bound to lead to a gradual accumulation of inflammable matter and eventually to an open *crisis* in the development of science and technology.[45] There is every indication that in the decades ahead the most advanced countries of the West will reach the threshold of this critical zone in the formation of the technological conditions of man's life.

An explanation has been advanced to the effect that it is exactly the *rapid progress* of science and technology and the excessively high rate of rationalization due to the scientific and technological revolution that destroy the life of man and society and 'create a sick social structure, a society that can no longer perform even its most basic functions in the accustomed way'.[46] However, analyses of the crisis phenomena present in the social aspect of scientific and technological development indicate that the problem lies rather in the opposite direction, in the inadequately dynamic utilization of the potential of science and technology, particularly in the stagnation of those areas of science and technology that, at a given stage of development of the productive forces, require priority (changes in the nature of work, protection of man and environment, etc.). All the present-day discrepancies are brought into distinct relief in the prognoses of future development. The prevalent conception in the West — namely, the 'transition from the past dynamics of growth to a future condition of world equilibrium'[47] — reflects the view that all these calculable alternatives of further development, based on an extrapolation of existing economic and scientific-technological growth, lead in their social implications towards a catastrophe — abstracting from or writing off the development of man and society.[48]

Just like the euphoria surrounding the first signs of the scientific and technological revolution, so the condemnation of science and technology following its first advances confuses *social content* with *technological phenomena*. The specific feature of capitalist society consists of the development of the basic class contradiction into abstract generality. The dominant class conditions have thus acquired the nature of general laws governing social life; they actually generated the abstraction of 'society as such' and the abstraction of the man-Nature relation as well.[49] The mode of existence of the class element in a general form, denounced by Rousseau as the divergence between phenomenon and essence in bourgeois society, was subsequently actualized in the relation of social and technological elements of production. It would seem that, within capitalist industrial production, where just technology (machinery) and manpower encounter each other, the social relation that founded this process of production is vanishing; that it is the all-human, eternal form of apparatus of man's interference with Nature that is left behind. This 'extinction' of the social form, however, is no more than an illusion relatable to the fact that the entire production process in its material form has been transformed into the production process of capital. The subordination of living labour to accumulated labour is actualized in all the material components of the production process, present both in the specific form of a machinery system[50] determining the course of production, and in the specific role of the labour force, in so far as it 'serves' technological systems. Moreover, the same social content is implicit in the forms and the trends of the sophisticated technology employed in the consumer system which remains the domain of the simple reproduction of labour force, underlying the extended reproduction of capital — no matter at what level the simple reproduction of labour force takes place.[51] The illusion of the extinction of the power of social relations in the world of industrial technology was spread already by Saint-Simon for postulating the contradiction between the 'industrial system' itself and its actual social agent, capital. The more distinctive and complicated the professional technological and organizational executive of capitalist production becomes with the growth of monopolies, the more the illusion actualized in technological expression of this social reality develops into the theory arguing that capitalism is dying out,[52] giving way to the 'industrial system'; that economic pressure is replaced by a mere 'Sachzwang'; that the control of the process of production by capital is replaced by the domination of the 'technostructure'; that exploitation has been reduced to its abstract outward form of alienation; that capitalist calculation has been replaced by an abstract form of rationalization.

Max Weber was still aware of the inner interrelation of this type of 'rationality' and the fundamental interest of capital.[53] In later interpretations, however, 'rationality' fell victim to 'real' illusion, thus becoming an abstract form, usually defined as deriving from the mere manipulation of things and a technological mode of controlling nature. 'Rationality' — in its empirical form pervading the life of capitalist society — entails its basic motivating core, i.e. the endeavour to 'attain with a given amount of wealth a maximum surplus product or surplus value',[54] in whatever way it may be corrected by rational considerations referring to the conditions where this endeavour could prove successful.[55] In this way, 'rationality' spontaneously reproduces its original social limitations, i.e. the conditions under which 'it is only by dint of the most extravagant waste of individual development that the development of the human race is at all safeguarded and maintained . . .'.[56]

In a bourgeois society, technological progress and subsequently science — in their present-day expansion — adhere to their own rational principles only in so far as, in these principles, the restricted 'rationality' is a priori incorporated. As a result, the development of science and technology follows a restricted one-sided course,[57] offering neither the feasible nor the optimal alternative of the scientific and technological revolution. On the contrary, within the ever-widening gamut of possibilities inherent in the development of the productive forces of social labour, this type of scientific and technological progress by-passes that major section which is connected with the all-round development of the potentialities and activities of the workers, thus suffering *structural deformations.*[58] In view of the mounting importance of the general development of human potentialities — under the conditions of the scientific and technological revolution — this structural disparity becomes more and more ominous. The adaptation of science and technology to the existing conditions is achieved at the cost of turning a considerable part of the productive forces of society into destructive forces. Understandably, social theory and social criticism, confusing social content with its technological expression, provide no answer as to whether and how society can shape the course of the scientific and technological revolution.

SOCIAL SYSTEMS AND THE DIRECTIONS AND BEARINGS OF THE SCIENTIFIC AND TECHNOLOGICAL REVOLUTION

The present advance of the scientific and technological revolution gradually pushes the earlier type of technological determinism (W.F. Ogburn) ad absurdam. Only those who are willing to risk the role of prophets of inevitable doom still adhere to this method in all its aspects and implications.[59] The gradual shift towards a more refined technological determinism and/or moral determination, towards conceptions envisaging technology as a 'dependent variable'[60] is apparent in the social theories of the West.[61] Similar difficulties — under the conditions of the scientific and technological revolution — beset the method based on cultural determinism and derived from Weber's concept. The clash between 'rationality' well developed in practice and its own after-effects either leads to a complete suspension of this principle and a new emphasis on the irrational — sought by modern 'enfants perdus' in the depth of a non-rationalized nature — or induces the endeavour to revise the original 'rationality' which is more and more exposed to the influence of moralizing subjectivism whose helplessness is evident in the suggestion to contain scientific and technological progress 'for the sake of man and society'. The result is either con-tinuous oscillation beween these two trends or increasingly sophisticated methodological combinations based on the 'reciprocal influence' of different factors. It is often frankly admitted that, on this basis, it has so far been impossible to specify the relation between accelerated scientific and technological progress and social change.[62]

The search for the clue to understanding the social roots of the scientific and technological revolution unattainable at the level of abstract technology-society interrelation, points to the *dialectics of productive forces and production relations.* No major changes in the structure of society can be effected without the mediating role of adequate social pro-ductive forces, i.e. without a corresponding 'revolution in productive forces, one that appears as a technological revolution'.[63] At the same time, all the major changes in the nature and level of productive forces which actually formed the new basis of production were connected with the operation of specific production relations as their adequate form of motion that carried them through.[64] Consequently, society cannot control the scientific and technological revolution until such social relations are introduced that allow for an adequate development of man and society. The scientific and technological revolution is thus essentially and intrinsically connected with revolutionary social changes.

Within the total set of possibilities specific to the present-day development of social productive forces — a set defined in terms of two different ways of linking up science and production (social labour) — the decisive factors are production relations. They determine which of the two possibilities is to become the reality and consequently what orientation the scientific and technological revolution will take. The course of the scientific and technological revolution is mediated through the entire mechanism of social life. Individual components of this process closely follow each other. Under suitable social conditions their consequences return to their causes, thus speeding up and intensifying the entire process. After completing this broad circle, the results attained provide a basis for further changes. Whereas for those social conditions that are not conducive to giving full rein to a number of components of the whole process and block their operation, the scientific and technological revolution remains confined to the narrow circle of the application of science and technology. Analyses lay bare two decisive links in this process of mediation, links that underlie not only a different type of social consequences of the scientific and technological revolution in the socialist countries but also a different type of the process itself:[65]

(1) As to its nature, science is a social productive force $\kappa\alpha\tau$' $\dot{\epsilon}\S o\chi\dot\eta\nu$. The socially combined process of labour provides a groundwork for the transformation of science into an immediate productive force and, vice versa, this use of science is connected with the process of integration in economic and cultural life. Social unification (socialization), social organization of the entire reproduction of human life is a prerequisite for the full appropriation of achievements of the scientific and technological revolution. In the process of this consumption, the results of scientific activity are not absorbed but improved and made available for further

progress. Science thus belongs to productive forces incompatible with any form of private ownership — monopoly of the means of production.[66] The elimination of boundaries imposed by private ownership of the means of production and utilization of science as a social productive force[67] are the first essential features peculiar to socialism and its attitude towards the scientific and technological revolution. The exclusion of all monopoly agents from this sphere — except for society itself — creates the possibility for the society to appropriate, as the active agent, the entire process of the scientific and technological revolution, to overcome particular interests not codified here by the superimposed private ownership system, and to establish a unified system of planning and control in the sphere of science and technology.[68]

In face of the widening scope of possibilities and necessities arising in connection with the scientific and technological revolution — the social functions of the entire system of planning and control in the sphere of science and technology are getting more intense and significant — this involves not so much the improvement of instruments essential for controlling nature, as the shaping of means of 'self-development' of man and society.[69] With 'the scientific and technological revolution providing the material groundwork for the development of social relations . . . the control of scientific and technological progress is one of the most powerful levers in the control of social relations within socialist society'.[70]

(2) The second difference pertinent to the scientific and technological revolution under socialism is the specific way of the development of socialist production relations in relation to man. As to its essence, science is in fact a form of *development of man's productive forces*,[71] of the potentialities of the social individual. Through its practical appli-

cation, science releases the potential for an all-round development of the abilities and creative activities of the working people. At the same time, the advance of science is essentially dependent on the development of man and his faculties. The development of production relations in a socialist society is geared towards a gradual concatenation of all the links within this circle. This, of course, presupposes comprehensive planning and systematic control of the complex social processes. [72] .It is not by chance that these issues gain in prominence with the advent of the scientific and technological revolution. On the one hand, the development of the socialist way of life and of socialist relations among people are the goals of scientific and technological progress. On the other hand, they are an active agent of this progress, ensuring the utilization of the released resources to attain the goals envisaged — i.e. continuous expansion of the life process of the working masses, the all-round development of their faculties — preventing wastage of this potential in fruitless or even harmful activities. The planned development and the controlled use of science, within the entire complex of its application, is orientated by the socialist society towards its major social goals — the general development of the potentialities of the working people. This implies an integration of the process of cognition and transformation of the world with the self-cognition and self-transformation of man and society. As a result a new type of science is emerging, consciously substantiating its advance by the knowledge of its own social prerequisites and aims directly relating to the ways and means of human life.

As the scientific and technological transformations and their implications in socialist society bring about the social development of the potentialities and activites of the working people, a new active element of the development of productive forces is being set in motion. This cause-effect,

ends-means dialectic opens up a dynamic system of interrelated changes in productive forces and production relations,
a *spiral-shaped* development[73] typical of the organic linkage
between the achievements of the scientific and technological
revolution and the social relations of an *advanced* socialist
society.

Empirical data, allowing for a comparison of different
attitudes towards the basic elements of the scientific and
technological revolution in different social systems displaying
the same level of labour productivity, testify to a much
higher degree of utilization of science and technology in
the socialist countries.[74] Per analogiam, some scholars
expected that this dimension of the use of science would
pose an increasing amount of social problems. In fact,
however, problems emerge in a completely opposite direction:
in socialist society there is an ever-growing need for further
expansion and acceleration of scientific and technological
progress. The point is that the development of productive
forces in socialist society (even where identical technological
components are employed) is taking place along a different
trajectory — namely, within the 'broader range' of the
appropriation of science and its applications by society as
the active agent mediated through the development of the
potentialities and activities of its members. At a certain stage
of the development of socialist society — under the conditions of the scientific and technological revolution — the
most effective way of multiplying the *productive forces*
of society inevitably proves to be the *general development of
man himself,* a social individual as an end in itself — the
growth of abilities and faculties of the worker.[75]

It emerges that rationality applied in the socialist system
and based on integration of the processes of transforming the
world and building up society belongs to a completely
different social force, operating in a different sphere and with

a different content. This type of rationality expresses the
endeavour of the working class to augment to the utmost
the wealth and the productive forces of society, but under
conditions in which the growth of wealth becomes increasingly dependent on the general development of man — to
exactly the same degree that the development of the
potentialities and activities of the working people affects
the development of productive forces, thus providing means
for the further development of man. It is only from this
process that man can develop as an end in itself. The coincidence of the development of productive forces and man's
social development offer after all the only way out of the
historical dilemma of goals and means.

Consequently, the *coincidence* of the development of
socialist production relations and the use of the achievements of the scientific and technological revolution[76]
represents the one and only road along which the general
development of man — i.e. not only of a class elite, but of
millions of people in all continents — can become a
social reality.[77] In this way, socialist production relations
develop into the absolute form of motion of the society's
productive forces.[78] In this light it becomes clear that
the road of growth of productive forces under capitalism
is narrow, conflict-ridden and restricted; that the 'rationality'
of capitalist calculation is divorced from the rational criterion
of maximum development of the productive forces of society
— development which has its indicators only in the general
system of the *economy of time* which is today systematically built up in socialist countries as an indispensable
guideline for complex social processes as a whole.[79] The
essential transformation of the content of labour, viewed
from the angle of traditional economic 'rationality', might
have appeared as a mere social measure, a basically un-
rewarding measure. Equally, the standard of education could

have been assessed from the restricted point of view of the need for qualified productive forces. A different picture emerges, however, when visualizing the scope and size of productive forces of technological and social progress, innovation activities, etc. placed at society's disposal by a goal-orientated transformation of the content of labour,[80] or by a systematic improvement of educational standards with regard to the impact of these measures on the development of man.[81]

Even in the social reality of today these changes called forth by socialist production relations are reflected in the fact that the same characteristic proportions specified for the field of scientific and technological activities in the socialist countries, likewise occur in all spheres related to the development of the capacities and abilities of the working people (e.g. education).[82] Understandably, from the point of view of traditional economic 'rationality' of the capitalist system, these phenomena appear as a deviation', as a social extreme, a chronic 'overinvestment of human resources', in the socialist countries.[83] In reality, these are all signs indicative of a different type of rationality, reflecting a specifically *socialist way of development.* Far from converging towards the structure and trends peculiar to technologically advanced capitalist countries,[84] this development steadily drifts away from them, projecting the contradictory social contents into *diverging trends* in the development of productive forces.[85] It is to be expected that with major scientific and technological changes in the forthcoming decade (such as the utilization of cybernetics or new discoveries in the field of biology) these divergent trends will increasingly prove to be a distinct form of contrary trajectories in the process of the scientific and technological revolution under way in the two antagonistic social systems.

SCIENTIFIC AND TECHNOLOGICAL REVOLUTION AND DEVELOPMENTS IN THE STRUCTURE OF SOCIETY

The central issue of sociological analysis of the scientific and technological revolution is the intriguing problem of the *prospect of social development,* of changes in the nature and structure of society under the conditions of the scientific and technological revolution. A feeling of temporariness is markedly spreading throughout the West. Individual symptoms of the process of the scientific and technological revolution, along with a number of other factors, provide subject matter for reflection upon the coming of a 'new society'; they underlie the advocacy of the advent of a 'new industrial', 'technological' or 'technetronic' society, 'post-industrial' society, 'post-traditional' or 'post-modern' era, a 'society of knowledge', 'learning society', 'informed society', 'civilization of services', 'tertiary' society, 'affluent' society, 'post-capitalist' or 'post-bourgeois' society, 'planetary' society and 'global' society — to quote but a few of the numerous projects to date.

Of the wide variety of schemes of this type, the majority refers to conceptions envisaging social roles specific to those social groups and classes that are believed to represent the guiding force of the scientific and technological revolution. A number of such conceptions proceed from changes in numerical proportions occurring on the surface of the complex social reality — directly subsuming these phenomena under the immediate consequences of the development of science and technology.[86] Their conclusions are primarily based not on a social analysis but on a technical-organizational classification of the functions of different social groups, in particular of the scientific and technological intelligentsia.[87] The confusion of production relations with their expressions in the technical and organizational structure — a phenomenon specific to advanced bourgeois relations — provides, in the early stages of the scientific and technological revolution, a framework for a number of technocratic conceptions of future social development, pushing back social theory to technological determinism while believing to have abandoned it.

This illusion behind these conceptions is that where power is armed with knowledge, it is eo ipso knowledge that controls power. Experience has shown that the scientific and technological revolution cannot immediately change production relations that provide a framework for its operation, neither can it eliminate the social, class barrier. There is another dimension in which the impact of the scientific and technological revolution is felt. Within a society based on antagonistic class-relations and giving no free rein to the formation of social productive forces inherent in the development of the scientific and technological revolution, there is also a *sharpening* of the existing antagonism to an extent that calls forth the need to rebuild the social foundations of life. While encountering non-antagonistic relations allowing for the development of all socially useful dimensions of science, and for involving the vast potential of the all-round development of the creative potentialities of the working people into the development of productive forces, the scientific and technological revolution offers an adequate *dynamic* basis for the development of society.

One of the vigorously debated questions is how far the two different types of utilization of the achievements of the scientific and technological revolution can go under differing social relations and systems, if these achievements themselves have so far been generally applied throughout the world with minor technical differences.[88] With this question, the concept of 'extinction' of social, class forms in their correspondingly adapted technological reality — as usually applied to the world of production relations — is being re-applied to the world of productive forces. The indentity of technical or organizational means, and equality in the general (quantitative) level of productive forces do not, of course, entail the identity of inner relations in the structure of productive forces and the trends of their development.

The general content of changes in the structure of productive forces connected with the scientific and technological revolution is the application of science as an immediate productive process. Due to different relations between science and the workers' collective, two different opposite forms of their linkage emerge under different social conditions in connection with the scientific and technological revolution. These two modes of linkage display different boundaries (the 'narrow' and the 'broad' range of the use of science) and are mediated through the different development of *social classes and groups,* through different configurations of interests, motivations, etc. The most decisive criterion for the different trajectories of the scientific and technological revolution from the viewpoint of the development of society is not only the scope of application of science but also the way of its utilization for the further development of the productive forces of society.

An essential characteristic feature of the capitalist mode of production is that it '. . . makes science a productive force distinct from labour and presses it into the service of capital'[8][9] thus using science *separately from labour* and in *opposition to labour.* Within the social structure of capitalist society, the initial stages of the scientific and technological revolution acquired the form of shifting proportions of some professional groups, particularly evident in the rapid growth of the group of scientists, engineers, technicians, etc.[90] engaged in the application of science in production; whereas for the workers, science remains an extraneous, alien sphere in which they practically do not participate and which is increasingly getting beyond their reach.[91]

Through the absolutization of the specific trends of the scientific and technological revolution and abstracting from capital — the decisive agent determining the particular use of science — the scientocratic and technocratic concepts of a new society, 'post-industrial society', etc.[92] were evolved, according to which the professional scientific elite becomes the leading force of society, with theoretical knowledge becoming its 'axial principle', and education ('intellectual property') providing the basis for 'stratification' or even 'class differentiation'. In essence, these conceptions (Bell) were designed to contend with the Marxist findings concerning the crucial role of the working class in our epoch.[93] They attribute to the process of the scientific and technological revolution the continuing fatal division 'between the scientific and technical classes and those who will stand outside',[94] i.e. those who will have to relinquish the prospect of their own development. This conclusion proves the untenability of such conceptions, as the consequences connected with this type of development are drastically at variance with the very prerequisites of the scientific and technological revolution itself.

The scientific and technical intelligentsia is no doubt interested in the development and application of science. This in itself, however, determines neither the basic social conditions (production relations) which provide the framework for the operation of scientific and technological progress, nor the specific orientation of science and technology towards definite ways and modes of development of the productive forces of society — since this interest can be pursued both in the 'narrow' and the 'broad' range of application of science as a productive force. In so far as the process of development and application of science is constrained by the operation of the antagonistic conditions of the extended reproduction of capital, in the position of the intelligentsia some elements of privilege are concealed with respect to the separation of the working class from science. Paradoxically, this diminishes to a certain extent the interest of the intelligentsia in the development of the productive forces of society (this development becoming increasingly dependent on the abilities and capacities of the workers) and attenuates — in the final analysis –- within the ranks of the intelligentsia the need for the development of its own creative potentialities.

On the other hand, of course, in capitalism under the conditions of the scientific and technological revolution, these elements of privilege diminish in an ever-increasing section of the professional intelligentsia as its numbers rapidly increase, thus enhancing its differentiation and bringing its substantial part closer to the other working people and, within this section of intelligentsia, inducing the urge to change the orientation and social intentions of science and its applications. At this point, the progression of the scientific and technological revolution under capitalism appears in a reversed form, as the relative overproduction of science and the professional intelligentsia. This is not because too much knowlege is being accumulated but because — against the background of the life limits of the working people — the powerful wealth of scientific knowledge and technological means serving the interests of capital clashes with its own base and social source, the development of the social individual, thus proving the absurdity of the goals pursued.

Following these lines, the scientific and technological revolution does not give birth to a 'new', 'post-industrial' society. On the contrary, the old capitalist social structure is working towards a deformation of the development of productive forces (including restrictions in the field of science and professional personnel).[95] Under these circumstances, when the manpower potential released from immediate production by the growth of productivity is shifting to the sphere of services, the prospects of the scientific and technological revolution are frequently interpreted in terms of the 'society of services', 'tertiary society', etc. and the hegemonic position of the 'new middle class'.[96] However, the growth of these spheres and social groups is ambiguous under capitalism;[97] with regard to the structure and dynamics of the productive forces, they stimulate the development of consumer motivation but in an antagonistic social form, wasting 'a world of productive capabilities and instincts' of the workers (Marx) and petrifying thus the original class barriers. All this points to an impasse in the development of society under conditions of the scientific and technological revolution, in so far as this process takes place within the antagonistic class structure and the corresponding 'narrow range' separating science from labour. For sociological analysis, this development represents a source of apprehension rather than a project for the future.

Another alternative, a *differing trajectory* of the scientific and technological revolution, is implicit in the structure of productive forces under the present conditions. Crystallizing with socialist production relations, this line aims at a completely different prospect of social development. The focal point lies in such a type of incorporation of science into production that is fully adequate to its function as a social productive force and *intellectual potential |of social labour;* in other words, it lies in the linkage of science with the main component and the most essential source of productive forces, the entire collective of the working people within the broad range of applications of science.[98]

The leading role throughout this process can only be played by that social force whose primary life interest is to break through and overcome all barriers between science and labour, since only in this way can the development of its own potentialities be given full rein. In the social structure of industrially advanced countries, there is only one such force, the *working class*,[99] since it is the only class whose fundamental life interest is to abolish all antagonistic social relations based on private exploitation of social labour (embodied in the respective material conditions) that restrict the worker's life in capitalist society, to overcome all the consequences of these relations, and to create both social and technical conditions for a life based on the general development of man's potentialities and powers. At the same time, the working class cannot ensure these conditions for its own sake without ensuring them for all working people, without carrying them through as a general law of social life. The more the working class asserts its own interests in the maximum development of productive forces, the more it unifies the interest of other groups of working people, particularly the professional intelligentsia, around the same goal, promoting within the ranks of these groups the interest in their own development. Thus the working class is gradually providing conditions under which every individual, through his own development, acts as a mediator of the development of all others, and vice versa. It becomes evident that it is exactly the *socialist and communist* relations that are adequate to the conditions of the scientific and technological revolution.

In the social structure of the socialist countries, the onset of the scientific and technological revolution is reflected not only in changes affecting the quantitative proportions of different social groups and classes,[100] but also, to the same degree, in the marked development and gradual 'self-transformation' within all social groups and classes,[101] their rapprochement under the leadership of the working class. In this way the scientific and technological revolution is promoted in a planned and systematic way to overcome the material sources of the traditional 'developmental contradiction' under which — as emphasized by Ricardo, Guizot or Hegel — the general development takes place at the cost of the development of the masses of individuals and vice versa.

The release of man from the immediate production process and the extension of his functions in the preparatory and control phases of production is likewise mediated here through a gradual restructuring of the activity of the entire working masses, involving the transformation of labour, its enrichment by creative scientific features,[102] the increasing participation in management, rationalization, innovation activities, etc. The labour force which is fully released by the scientific and technological revolution from the immediate production process is systematically transferred into areas decisive for linking science with man's over-all development.[103] The conditions of the scientific and technological revolution have thus not in the least detracted from the leading role of the working class in society, determined by its position in the system of production relations and in the social organization of labour, actualized under the guidance of the party, constituting this class as the main active agent of social development and linking it, at the same time, with scientific knowledge. Each advance of the scientific and technological revolution enhances the role of the working class — the *crucial social force of our epoch* — as the only class capable of paving the way, on the basis of the process of revolutionary changes in production relations, towards the universal development of the productive forces, conducive to man's all-round social development.

Within the complex of its social aspects, the scientific and technological revolution emerges as a change in the structure and dynamics of the productive forces corresponding to the nature of our epoch, the revolutionary transition from capitalism to communism; essentially linked with the regularities of this process, it creates for this new society an adequate material technological basis. The prospects and possibilities that the scientific and technological revolution offers for the development of society round off the definition of its nature. In conclusion, I regard it as my duty to state that international cooperation in considering these questions is only possible if scientific inquiry is not perverted into political invectives. We have therefore to refuse resolutely the regrettable attempt to push through into scientific discussion unsubstantiated political invectives against the activities of Soviet leaders which found its clear expression in the contribution of Professor Touraine.

NOTES

1. '.... the United States has become the technological society ...', J.D. Douglas, Preface to: *Freedom and Tyranny, Social Problems in a Technological Society* (New York, 1970) vii.
2. Cf. M.G. Mesthene, *Technological Change. Its Impact on Man and Society* (Cambridge, 1970) 15.
3. R.K. Merton, *The Sociology of Science* (Chicago-London, 1973) 175; B. Barnes, *Sociology of Science* (Harmondsworth, 1972) 11.
4. Labelling the term 'scientific and technological revolution' as unsubstantiated and premature, the Report *Technology and the American Economy* Vol. I (Washington, February 1966) 1-2, accentuated the conception of 'continuous technological change'.
5. G.T. Seaborg and W.R. Corliss, *Man and Atom*, (New York, 1971).

6. A.M. Weinberg, *Reflexions on Big Science* (Cambridge-London, 1967) 6.

7. J. Diebold, *Man and Computer. Technology as an Agent of Social Change* (New York, 1969) 23.

8. Cf. Bernal's studies *Science in History* (1954), *World Without War* (1958), etc.

9. D.M. Gvishiani, Nauchno-tekhnicheskaya revolyutsiya i socialniy progress, *The Symposium of Scientists and Specialists from Countries of the CMEA* (Moscow, 1974). Gvishiani describes the specification of the 'nature and the social role' of the scientific and technological revolution as 'one of the most significant theoretical and political findings of Marxism in the second half of the twentieth century'.

10. Cf., for example, G.N. Volkov, *Chelovek i nauchnotekhnicheskaya revolyutsiya* (Moscow, 1972); I.A. Mayzel, *Nauka, avtomatizatsiya, obschchestvo* (Leningrad, 1972) and other publications.

11. Cf., for example, J.S. Meleschchenko, *Tekhnika i zakonomernosti yego razvitiya* (Leningrad, 1970); S.S. Tovmasyan, *Kachestvennye fazy razvitiya tekhniki i sovremennaya nauchno-tekhnicheskaya revolyutsiya* (Erevan, 1970).

12. The dialectics of this issue is the subject of B.M. Kedrov's (cf., for example *Nauchno-tekhnicheskaya revolyutsiya i socialism,* Moscow, 1973) detailed criticism of the simplistic interpretations of this process subordinating the entire social development to science.

13. K. Marx, *Grundrisse der Kritik der politischen Ökonomie* (Berlin, 1953) 594 and 588.

14. V.I. Lenin, *Collected Works,* Vol. 1 (Moscow, 1963) 85.

15. Cf., the Soviet-Czechoslovak publication *Man-Science-Technology. A Marxist Analysis of the Scientific and Technological Revolution* (Prague, 1973) 364. Cf. also N.N. Markov, *Nauchnotekhnicheskaya revolyutsiya: analiz, perspektivy, posledstviya* (Moscow, 1973) 11.

16. The category of productive forces is frequently misinterpreted. J. Ellul, *La technique et l'enjeu du siècle* (Paris, 1954); D. Bell, *The Coming of Post-Industrial Society* (New York, 1973) as well as a number of other writers identify the Marxist concept of productive forces with technology — thus excluding the relevant social and

human components from its domain.

17. Cf. the entry 'l'industrie' in the *French Encyclopaedia*, by de Jaucourt.

18. K. Hager, *Socialismus und wissenschaftlich-technische Revolution* (Berlin, 1972) 23.

19. 'Discussing science as a source of the development of productive forces, we bear in mind both the immediate improvement of objective factors of production and the man as a subjective productive force', V.G. Marachov, *Struktura i razvitiye proizyoditelnikh sil socilaisti-cheskogo obshchestva* (Moscow, 1970) 157-58.

20. 'The accumulation of knowledge and skill, the general productive forces of the social brain is thus, contrary to the labour, absorbed in capital, thus appearing as a quality fo the capital, more specifically, the quality of *capital fixe* . . . ', K. Marx, *Grundrisse der Kritik der politischen Ökonomie* (Berlin, 1953) 586.

21. In Weber's well-known metaphorical description of the machine the contradictory nature of this use of science is inherent: 'A machine is "coagulated intellect" (geronnener Geist). It is precisely this fact that gives science the power to compel people to its office, the power to dominate and determine their working day in their day-to-day life, as is the practice in a factory', *Gesammelte politische Schriften* (München, 1921) 151. Enlarging upon Weber a conception of science — contrasting with Humboldt's approach which associates science with man — J. Habermas bases present-day science on the domination-legitimizing function, *Technik und Wissenschaft als 'Ideologie'* (Frankfurt a.M.) 92.

22. 'In this transformation it is . . . the appropriation of his [i.e. man's] own general productive force, his understanding of nature and his mastery of nature through his existence as a social body — in a word the development of the social individual appearing as a major fundamental pillar of production and wealth', K. Marx, *Grundrisse der Kritik der politischen Ökonomie* (Berlin, 1953) p. 593.

23. Ibid., 599.

24. R.K. Merton, 'Social and cultural context of science 1970'; in *The Sociology of Science* (Chicago-London, 1973) 173-75.

25. Cf. J. Ben-David, *The Scientist's Role in Society. A Comparative Study* (Englewood Cliffs, 1971) 4.

26. The claim that the 'Machbarkeit' of the world (H. Schelsky) is the brain-child of 'technological intellect' and a symptom of 'technocratic practice' is justified only in a situation under capitalism, where the strait jacket of the 'narrow range' is strictly imposed on science.

27. A.W. Gouldner holds that some of the sociological theories 'may be expected to have difficulty' in adapting to the consequences of technological progress, *The Coming Crisis of Western Sociology* (New York, 1970) 344.

28. Cf. R.K. Merton, *The Sociology of Science* (Chicago-London, 1973), 176-77.

29. S.N. Eisenstadt, *Tradition, Change and Modernity* (New York-London-Sydney-Toronto, 1973) 237.

30. H. Brooks (Ed.), *Science, Growth and Society* (OECD Report, Paris, 1971) 19 and 26.

31. E. Fromm, *The Revolution of Hope* (New York-Evanston-London, 1968) 3.

32. C.R. Taylor, *Rethink. A Paraprimitive Solution* (London, 1972) 129.

33. I. Illich *Celebration of Awareness* (New York, 1969).

34. R. Rozsak, *The Making of a Counter-Culture* (London, 1970).

35. As early as the mid-1960s, H. Ozbekhan comments (in two SDC studies: *Technology and Man's Future; The Triumph of Technology*): 'can' implies 'ought' (on the first signs of declining enthusiasm and the doubts looming large over technological progress): 'it is all, by now, probably beyond our control. As Pascal put it: Once embarked, il faut parier!'

36. The National Academy of Sciences' Report, *Technology; Processes of Assessment and Choice* (Washington, 1969) 1, points to the apprehension caused by the society being unable 'to channel technological developments in directions that sufficiently respect the broad range of human needs'.

37. 'The technological utopia that has guided the American dream seems to have been grotesquely inverted', W. Kuhn, *The Post-Industrial Prophets* (New York, 1971) 2.

38. A. Etzioni, *The Active Society* (New York, 1968) vii.

39. A.W. Gouldner, *For Sociology, Renewal and Critique in*

Sociology Today (New York, 1973) 76.

40. Cf. for example, M.W. Thring, *Man, Machines and Tomorrow* (London, 1973).

41. 'The majority of adults in this country hate their work. Whether it is a factory job, a white-collar job, or, with some exceptions, a professional job . . .', C.A. Reich, *The Greening of America* (New York, 1970) 274.

42. Thus Di Palma has drawn attention to the growth of social inquality attendant upon specific types of scientific and technological progress in capitalist countries (*Apathy and Participation, Mass Politics in Western Society,* New York, 1970).

43. D. Riesman, 'Lesiure and work in post-industrial society', in *Mass Leisure,* (Glencoe, 1958) 379.

44. *Prospect for Mankind* (Hudson Institute, 1972) 59.

45. A number of prognoses single out the mid-1980s as the scene of the final flare-up of this 'technological crisis', H. Kahn and B. Bruce-Briggs, *Things to Come. Thinking about the Seventies and Eighties* (New York, 1972) 205 a.o.

46. A. Toffler, *Future Shock* (New York, 1971) 185.

47. J.W. Forrester, *World Dynamics* (Cambridge 1971) 2.

48. Cf. D.H. Meadows, D.L. Meadows, J. Randers and W.W. Behrens, *The Limits to Growth* (New York, 1972) 46.

49. Ancient Rome and Greece knew the community and the state, the Middle Ages the estate and the country; 'society as such' emerges only with the bourgeois society. 'Thus capital first creates a bourgeois society and the universal approximation of nature and the social bond itself, through the members of society', K. Marx, *Grundrisse der Kritik der politischen Ökonomie* (Berlin, 1953) 313.

50. 'Along with the machine, the domination of past labour over living labour becomes not only a social fact expressed in the capitalist-worker relation but also — so to say — a technological reality', K. Marx, unpublished papers, quoted after *Voprosy istorii yestestvoznaniya i tekhniki* (25/1968) 74.

51. 'Means of life buy the worker so as to incorporate him into the means of production. Means of life are a special material form of existence in which capital is opposed to the worker', *Archiv Marksa i*

Engelsa, Vols. II, VII (Moscow, 1933) 234-36.

52. J. Schumpeter, *Capitalism, Socialism and Democracy* (New York, 1942).

53. 'The rationalization of economy today is an inquiry into a rational organization of economy from the point of view of capital interest', M. Weber, *Die Fragen der Rationalisierung,* (Zürich, S.A.) 9. The 'cultural' criticism of technological reality sidetracking the social roots of Max Weber's conception of 'rationality' gets no further than the repeated diatribe against 'technological rationality' based on the belief that 'technology per se' is domination over society and nature alike, H. Marcuse, 'Industrialisierung und Kapitalismus im Werk Max Webers', *Kultur und Gesellschaft,* Vol II (Frankfurt a.M. 1965) 127. Even when attempting to progress further to discern the operation of bureaucracy as the source of 'technological reason', this criticism gets into a vicious circle since, according to Weber, it is precisely 'technological reason' that gave rise to bureacracy.

54. K. Marx, *Theorienüber den Mehrwert,* Vol. 2 (Berlin, 1959) 563.

55. During the discussion on 'technology assessment' B. Wynn pointed out the fact that 'discussed in a political vacuum, technology assessment is merely in danger of being used to refine and prolong the existence of inequality, exploitation and inhumanity . . . ', *Technology Assessment and Quality of Life* (Amsterdam-London-New York, 1973) 286-87.

56. K. Marx, *Capital,* Vol. III (Moscow: Progress Publishers, 1971) 88.

57. It should be pointed out that the prognoses of socialist scholarship which, as early as the inception of the scientific and technological revolution drew attention to the danger arising from the orientation of science and technology in the West, proved to be correct. Cf., for example, G.V. Osipov, *Tekhnika i obschchestvenniy progress* (Moscow, 1959) 63-64; similarly A.A. Zvorykin and others.

58. Cf., for example, *Harmonizing Technological Development and Social Policy in America* (The American Academy of Political and Social Science, Philadelphia, 1970). American literature frequently refers to the fact that the R & D pattern in the US 'was badly skewed', D. Bell,

The Coming of Post-Industrial Society (New York, 1973) 260.

59. Cf., Z. Brzezinski, *Between Two Ages. America's Role in the Technetronic Era* (New York, 1970) 9.

60. F. Hetman, in *Society and the Assessment of Technology* (OECD, Paris, 1973) 389.

61. Cf., G. Friedman, *La puissance et la sagesse* (Paris, 1970).

62. 'Only with far more research in hand than has yet been done' is it possible to answer the question of 'whether, how, and to what extent society is shaped by its technology' (Harvard University Program on Technology and Society, 1964-1972, *A Final Review*, Cambridge, 7).

63. K. Marx, unpublished papers; quoted after *Voprosy istorii yestestvoznaniya i tekhniki* (25/1968) 51.

64. The replacement of the dialectics of productive forces and production relations by a 'more abstract' relation of 'labour and interaction' — proposed by J. Habermas, *Technik und Wissenschaft als 'Ideologie'* Frankfurt a.M., 1968) 92 — is again equal to a reduction of social processes to their outer material form.

65. Naturally, the analysis of differences in the operation of social systems in relation to the scientific and technological revolution requires the exclusion of differences in the general level of productive forces (labour productivity); e.g. the present approach to science, technology and social progress in the Soviet Union as compared to the United States in the immediate post-war years. (The twenty-five-year-long distance in the temporal scale of the US is, of course, equivalent to the much shorter distance in the temporal scale of the USSR, a country which, due to its considerable advance in production relations, has within the last ten years undergone a period which took about 50 years in the US.)

66. Due to this 'communist' feature of science (cf. R.K. Merton, *Social Theory and Social Structure*), private ownership must necessarily appear to the society using science as an immediate productive force — similar to ancient Greece and Rome — as 'privation' (privatus — bereft, excluded from the disposition right of the community). Attempts to define 'knowledge' in terms of 'intellectual property', cf., for example, D. Bell, in *Indicators of Social Change. Concepts and Measurements,*

(New York, 1968) are concomitant to confining science to the activity of a special elite and to the boundaries of the 'narrow range' of its application. In this way, new social phenomena are subsumed under old categories — just as King Arthur's Court was unable to envisage the Yankee republic in any other way than as a state composed of monarchs.

67. '. . . in order to build up communism, we must take technology and science and make them available to wide circles . . .', V.I. Lenin, *Collected Works*, Vol. 30 (Moscow) 458.

68. It was Lenin's idea that inspired the practice of planning and control of science. As late as the 1930s Bernal's summary of principles for the planning and control of science met with vigorous protest, as if this represented an attempt to 'deprive science of its freedom'.

69. In the Soviet Union during the 1960s, the number of natural scientists in research more than trebled. Characteristically, the number of specialists in the main branches of social sciences was increasing at a more rapid rate. Heading the list are: Economy (4.1 times), Philosophy and Sociology (3.6 times), Physics and Mathematics (3.3 times), Technical Sciences (3.2 times) (cf. D.M. Gvishiani, S.R. Mikulinsky, G.A. Kugel, *Nauchno-tekhnicheskaya revolyutsia i izmeneniya struktury nauchnych kadrov v SSSR*. Moscow, 1973, 69). In the Soviet Union, the proportion of social sciences is about 21 per cent, while in the United States the respective estimated figure is 17-18 per cent.

70. V.G. Afanasyev, *Nauchno-tekhnicheskaya revolyutsiya, upravleniye, obrazovaniye*, (Moscow, 1972) 187. This approach is essentially different from the practice adopted in the capitalist countries where 'technology is not harvested', D.N. Michael, *The Unprepared Society* (New York-Evanston-London, 1968) 38, and which hence 'are ill-equipped governmentally to plan and use such technological developments to achieve social aims', H. Perloff, 'Toward the Year 2000. Work in Progress', *Daedalus* (Summer 1967) 681.

71. K. Marx, *Grundrisse der Kritik der politischen Ökonomie* (Berlin, 1953) 439.

72. Social planning and control in the socialist countries is based on planning and controlling the whole systems of productive forces

and production relations; the sphere of production is seen as a sphere where not only objects but first and foremost human life is produced. Hence the tendency to build modern production units as bodies of complex economic, scientific-technological and social planning and control.
73. J. Filipec and R. Richta, *Vědeckotechnická revoluce a socialismus* (Prague, 1972).
74. The figures for 1969 give the follwing information on the relative intensity of scientific and technological activity (per capita R & D activities in relation to GDP per capita).

TABLE A

Groups of countries	According to relative share of scientists in population	According to relative level of expenditure on R & D from GNP
9 capitalist countries	86	85
USA	100	100
7 socialist countries	249	185

(Based on UNO, UNESCO, OECD data; the ECE method of comparison was adopted; 9 capitalist countries: Belgium, France, Holland, Italy, Japan, Norway, FGR, Sweden, Great Britain; 7 socialist countries: USSR, Bulgaria, Czechoslovakia, Hungary, GDR, Poland, Rumania.)

75. Marx foresaw that in the future society, the development of the abilities of the working people will, as a powerful productive force, shape in turn the development of production, providing the criterion of wealth. The present stage of socialist construction, requiring the application of the achievements of scientific and technological revolution, gives special significance to Marx's words. In an advanced socialist society, man's problems cannot but represent the first priority', G.E.

Glezerman, in: *Ekonomicheskiye i socialno-politicheskiye problemi kommunisticheskogo stroitelstva v SSR* (Moscow, 1972) 287-88.

76. In the practice of the socialist countries, this historical process acquires the form of the 'organic linkage of the scientific and technological revolution with the advantages of the socialist economic system', L.I. Brezhnev, *Leninskim kursom, Rechi i stati,* Vol. 3 (Moscow 1972) 257.

77. The development of science and technology is one of the constitutive features of this process. Attempts to present the endeavour of the Soviet Union and other socialist countries, aiming at the acquisition and further development of the achievements of the scientific and technological revolution, as a sign indicative of a certain degree of 'technocratization' of the socialist society — whether this takes the form of romantic conceptions, glorifying the 'intermediate', 'soft' technology (Schumacher) and/or the diverse forms of Maoist propaganda, which cultivates these and similar petty bourgeois contentions — are theoretically based on an approach that totally ignores the crucial significance of science as the 'essential form of human wealth' (Marx) indispensable for man's development and channels the development of society into a deadlock.

78. Socialist society 'is not based on the development of productive forces, a development that would reproduce a definite state or, at the most, extend this state, but where the free, uncurbed, progressive and universal development of productive forces becomes the prerequisite of the society', K. Marx, *Grundrisse der Kritik der politischen Ökonomie,* (Berlin, 1953) 438.

79. '. . . the scientific and technological revolution and the control over social processes nowadays represent the most essential means of the economy of time', V.G. Afanasyev, *Nauchno-technicheskaya revolyutsia, unpravleniye, obrazovaniye* (Moscow, 1972) 25.

80. Cf. I.I. Changli, *Trud. Sociologicheskiye aspekti teorii i metodologii issledovaniya* (Moscow, 1973).

81. Cf., V.N. Turchenko, *Nauchno-technicheskaya revolyutsiya i revolyutsiya v obrazovanii* (Moscow, 1973).

82. The figures for 1969 give the folowing information on the mobilization of 'human resources' in relation to GDP per capita:

TABLE B

Groups of countries	According to relative share of university students in population	According to relative share of teachers in population	According to relative level of expenditure on education from GNP
9 capitalist countries	68	111	91
USA	100	100	100
7 socialist countries	146	218	165

(Based on the same statistics and methods as Table A.)

83. F. Harbison and C.A. Myers, *Education, Manpower and Economic Growth* (New York-Toronto-London, 1964), etc.

84. Lenin's idea to 'catch up' with and 'overtake' the most advanced capitalist countries where standards of labour productivity are concerned has never been understood in the sense of an attempt geared towards the reproduction of social patterns and values of capitalism but on the contrary as a means enabling socialist society to materialize its own social aims and values.

85. The specific orientation of scientific and technological development on social aims is likewise reflected in the technical assistance extended by the socialist countries to the developing countries. This is, of course, a principle that is totally different from the drive toward maximum monopoly profits, which governed the practical policies and strategies of the former colonial powers vis-à-vis the subjugated nations and which even today pervades the technical aspects of policy pursued by industrially advanced capitalist countries. From a sociological point of view, it is therefore totally untenable to equalize these two phenomena — mutually exclusive as to their social content — under the vague label of categories such as 'West-East', 'North-South', 'forerunners-latecomers', etc.

86. Cf., for example, K. Boulding, *The Meaning of the Twentieth Century* (London, 1965) 171.

87. P.F. Drucker, *The Future of Industrial Man* (New York, 1965) 148.

88. 'To what extent can production relations and social organization vary under the same productive forces, in a situation where productive forces, science and technology are becoming more and more identical in all advanced societies?' R. Aron, *Les désillusions du progrès* (Paris, 1969) 251.

89. K. Marx, *Capital* (Moscow, 1965) 361. '. . . intelligence in production expands in one direction because it vanishes in many others. What is lost by the detail labourers is concentrated in the capital that employs them', ibid.

90. The technological and scientific revolution in productive forces broadened the masses of the working intelligentsia . . .', V.S. Semyonov, *Kapitalizm i klassi* (Moscow, 1969) 351.

91. Cf., for example, the disproportion between the 50 per cent increase in the number of scientifically trained specialists (professionals technical workers, etc.) in the US in the 1960s (i.e. up to 14 per cent of the active population) and the almost constant role of science in the training and activity of the workers (during this period, their number stabilized on 35 per cent of the active population).

92. 'To some extent this is an old technocratic dream which one can find anticipated in extraordinary fashion by . . . Henri de Saint-Simon', D. Bell, 'The measurement of knowledge and technology', in: *Indicators of Social Change* (New York, 1968) 238. V.I. Lenin pointed to the fact that the revival of Saint-Simon's ideas in the twentieth century is in close connection with the advance of monopolism.

93. D. Bell uses his concept of contemporary scientific and technological progress as an argument against the crucial role of the working class overcoming the barrier of capitalism in the process of founding and developing the socialist society. Attempts to interpret in this context some conclusions of our research into the nature of the scientific and technological revolution are at variance with the spirit and letter of our inquiry.

94. *The Coming of Post-Industrial Society* (New York, 1973) 112. The conclusion is that '. . . groups which lose out early in the educational race will be quickly excluded from society as a whole', 'The post-industrial society', in: E. Ginzberg, ed., *Technology and Social Change* (New York-London, 1964) 50.

95. Over the past three years, the proportion of professionals in the active population in the US has stagnated. Statistical data differ from the expected trends outlined in the 1960s (cf., for example *Technology and the American Economy,* Washington, 1966, 30)

96. Sociological studies reveal with increasing clarity the inner heterogeneity of these social groups, thus laying bare the futility of attempts to include among them skilled workers (cf. T.B. Bottomore in: *Contemporary Europe: Class, Status and Power,* London 1971).

97. If we isolate the section A within the sphere of 'services' (health care, education, social security, culture and science) and compare it with other areas of 'services' (trade, administration, finance, etc.), we obtain the following data for 1972:

Proportion of section A (in %)	USA	USSR
— in the active population	14	16
— in the total area of 'services' excluding transport	29	55

(Based on national statistics and the *Year Book of Labour Statistics,* Geneva, 1973).

98. G.N. Volkov, *Sociologiya nauki* (Moscow, 1968) 179.

99. Marx and Engels based the definition of the crucial role of the working class on the fact that workers 'are able to carry through their complete — no longer restricted — active self-assertion consisting in the appropriation of the totality of productive forces and in the development of the totality of abilities given by these productive forces', K. Marx and F. Engels, German ideology', in: K. Marx and F. Engels, *Werke,* Vol. 3 (Berlin, 1962) 68.

100. However rapid was the growth in the number of specialists in the Soviet Union (the 1960s brought a 90 per cent increase), the

bonds between science and the working class in no way lagged behind. (Cf. V.I. Bolgov, *Byudget vremeni pri socializme,* Moscow, 1973.)

101. Indicative of this process is the growing group of workers with higher education, (Cf., for example, M.N. Rutkevich and F.F. Filippov, *Socialniye peremeschcheniya,* Moscow, 1970, 120).

102. The Polish-Soviet study, *Socialniye problemi truda i proizvodstva* (Moscow-Warsaw, 1969, 49-50), indicates that, in the value system of workers in the socialist countries, creative work is heading the list of priorities.

103. P.N. Fedoseyev, *Kommunizm i filozofiya* (Moscow, 1971) 429.

3

OBSERVATIONS ON SCIENCE AND TECHNOLOGY IN A CHANGING SOCIO-ECONOMIC CLIMATE

Ralf Dahrendorf
London School of Economics

In the nine countries of the European Community in which I had the portfolio for Research, Science and Education, about 14 billion Units of Account (or 17 billion dollars) were spent on research and development (R & D) in 1973. Private agencies, mostly industry, account for just over half of this sum; the rest was public expenditure, amounting to more than 7 percent of all public budgets, national and regional, in the nine countries. The annual increase of overall R & D expenditure has been in the vicinity of 10 per cent for some time, although it was consistently somewhat lower after 1967 than before (especially in real terms). There is no overall plan for spending R & D funds in the Community. A committee has recently been set up to pave the way for co-ordination (CREST); however, this committee finds a great deal of uncertainty about priorities of research, methods of providing incentives, and channels of application in the member states as well. It is widely assumed that research will be useful in terms of aiding economic growth, providing basic services, and guaranteeing military security. From this it follows that science and technology as objects of public interest are, at least in terms of the dimensions of expenditure, essentially physical science and engineering; a small shift to the social sciences and a rather larger one to medical research has recently been observed.[1]

Such are the facts with which public figures are expected to impress their audiences. I am sure you are not impressed; yet the facts provide a useful starting point for four related observations.

The first of these observations is polemical and thus probably a little unjust. Seven percent of public expenditure is a large portion of the cake (in the United States and Japan the percentage is even bigger) and therefore indicates a considerable public interest; this is underlined by the proximity of research priorities to the core values of the societies in question. Thus one might well feel inclined to accept Daniel Bell's claim that we are in the process of moving from an 'industrial society' preoccupied with the production of goods to a 'post-industrial society' preoccupied with the production of goods to a 'post-industrial society' organized around knowledge which not only gives the 'professional and technical class' a pre-eminent place in society but also establishes the special claims of science against other social forces:

> The defence of science — against bureaucratization, against political subjugation, against totalitarianism — derives eventually thus from the vitality of its ethos. The charismatic aspect of science gives it its 'sacred' quality as a way of life for its members. Like Christianity, this charismatic dimension has within it a recurrent utopian and even messianic appeal. It is the tension between those charismatic elements and the realities of large-scale organization that will frame the political realities of science in the post-industrial society.[2]

But then, although I can see that the large-scale organizations of physicists and physiologists and perhaps even psychologists agree — what is the vitality of the scientific ethos without the funds to pursue research? It is true, when the Committee on Science and Technology of the German Federal Parliament recently invited leading scientists to testify about desirable priorities, they established a list which almost turned that of actual priorities upside down and thereby demonstrated their independence; but then, these same scientists are the main recipients of public money to conduct the research which they deprecated in public testimony. The first point to be made here is that research and development, far from transcending the values of the society surrounding it, is in fact part of it, an indispensable element of an existing socio-economic structure which is oriented towards the production of goods and services, that is, extensive growth. Indeed, in the terms of Max Weber, it could be argued that science and technology are the epitome of a society which has made instrumental rationality its guiding value.

But of course this is only part of the truth, and here I have to redress my injustice towards Daniel Bell as well as the scientists testifying to the German parliamentary committee. Science and technology are part of the prevailing structure of socio-economic relations in an industrial society, capitalist or otherwise, but they are at the same time a motive force in the development of a new potential for satisfying human life chances. Using the language of Karl Marx (which I believe is unnecessarily restrictive in its emphasis on material production), science is both an element of the relations of production and a powerful force of production. Possibly, the charismatic elements of science as an activity have something to do with this dual position. Certainly, the practical ambiguity of scientific theory has a story to tell: for quite often, the uses of scientific research are unknown when it is begun; apparently highly theoretical and irrelevant findings can turn out to be directly applicable to topical problems; which means that the impact of research findings can be unexpected in its direction and unwanted by its sponsors. The very fact that science thrives on a medium of uncertainty makes it a doubtful ally of any status quo.

These are not abstract remarks. For my second observation is that we are living through a period of important changes in the socio-economic climate, at least for the advanced societies of the world, and that science and technology have a key role to play in this process. There is many an indication that the institutions created for two centuries of extensive growth are no longer capable of coping with the potential for satisfying human life chances developed within them: concern with environmental problems, the so-called 'energy crisis', the headache and charm of our cities, the separation of education from its economic purpose and, last but not least, the persistence of the 'new inflation' as a caricature of growth, expansion without substance, are an almost miscellaneous collection of significant symptoms pointing to a change of the theme of historical development. Instead of expansion, we may be moving into a period of intensive growth, of conservation and containment on the one hand, of the rearrangement of economic and social patterns to enhance the quality of individual lives on the other.

There is a romantic approach to the problems raised by such transformations. It is apparent in the attitudes and behavior of some of the young who are tending to opt out and follow Rousseau's old battle-cry 'Back to Nature!'; and it is reflected in political analysis and action from Sicco Mansholt to Ivan Illich. If one wants to return to a partly pre-technical mode of life, science and technology clearly lose their central role, but so do liberty and welfare, for the return to Rousseau usually ends up with Hobbes, and the politics of cultural despair has always been illiberal and destructive. When I speak about intensive growth I am thinking of an improvement of individual life chances which, if anything, is more dependent on scientific research and its applications, as well as growth of a certain kind, and which moreover implies and produces respect for the autonomy of that great and representative human activity which we call science.

To make such statements specific, my third observation is about the kind of problem which would deserve priority in a socio-economic climate of intensive rather than extensive growth. At all times, mankind had to tackle two sets of issues, those of survival, and those of justice. Two centuries of extensive expansion have increased the scale of our problems of survival to the point where some eminent scholars believe that mankind is more likely to wipe itself off the face of the earth in the course of the next century than to survive.[3] Even if one is not afflicted by the doomsday syndrome to the same extent,[4] one cannot fail to see at least four major challenges which require massive research as well as political action:

1. The population challenge: the acceleration of medical, pharmaceutical and psychological research to facilitate birth control is one side of the picture; the other, equally important one, is an enormous effort in agricultural research and technology transfer and of course a solution for the development problems of the third and fourth worlds.

2. The energy challenge: we have entered the nuclear age with all its benefits and dangers; but these dangers as well as the limits of energy availability and usability demand a rapid increase of research efforts into post-fission nuclear resources and above all non-conventional, renewable

sources of energy.

3. The environment challenge: the change in social temper began with recognition of this problem, but our efforts must not slacken; science and technology can above all aid in the development of economical methods of environmental protection.

4. The nuclear challenge: here I am thinking of the hazards of the nuclear age, military and non-military. Reactor safety, nuclear waste disposal, but above all the geography and economy of the exploitations of nuclear resources continue to pose major problems of research and of the scientific imagination.

The problems of justice, of the right social, economic and political order, are no less intricate and often related to those of survival. Their study requires much work and imagination by social scientists: on the problems accompanying an economy no longer geared to extensive growth; on the international economic and political framework within which the welfare of all countries is best served; on the re-arrangement of education, work and leisure in advanced societies; on the political structures adequate for a period of intensive growth; in short, on the problems of the Mature Society (to use Dennis Gabor's expresssion).[5]

This list of priorities is indicative rather than exhaustive. In fact, many of the subjects mentioned already occupy a prominent place in the research budgets of institutes, public and private. Nevertheless, a re-orientation as I have indicated it might well affect, in the case of the European Community, about one-half of the total research expenditure and reduce many a current priority to second or third place: supersonic transport, the development of new products which are adding quantity rather than quality to the market, competitive systems of nuclear reactors, or or uranium enrichment methods or of colour television sets, etc. Needless to say, a fundamental re-orientation would require the supplementation of Strategic Arms Limitation Talks by agreements to limit research deliberately aimed at increasing the deadliness of weapons, although I admit that such research is more difficult to control, and to detect, than nuclear explosions in the atmosphere, or underground, and that there is always something dubious about the limitation of research.

I have talked about the order of magnitude, the place, and the subjects of science and technology at a time at which these social forces occupy a pivotal position between existing social structures and the potential for their transformation. This leads naturally to my fourth and final observation which is about science and politics. This is a traumatic relationship in almost every respect I can think of. I have mentioned the paradox of actual and desired priorities which reflects the differing values of politicians and scientists. The relationship between the scientific and the political communities is often uneasy and wrought with friction. In all countries with which I am familiar, the formal links between governmental structures and scientific research present unresolved problems of organization and communication. Parliamentary and public debates about the vexing subject of 'technology assessment' are often confused and confusing. The theoretically all-important subject of technology transfer (to developing countries, for example) is in practice little more than a word. Indeed, there is a growing suspicion that we are still far from having found satisfactory methods for feeding scientific research into the political process, especially in the social sciences, while (or perhaps, because?) politicians continue to feel that their concerns are insufficiently recognized by scientists. Perhaps my own experience is untypical, or European countries are particularly backward in this respect, but I suspect that the problems exist in other areas as well.

Organizational difficulties in the relations between governments and science are aggravated by the fact that the position of scientists as a social category is no less ambiguous in the advanced societies today than that of science as an institution. One of the important social developments of the last decades is the increase in the number and proportion of adults (18 years of age and more) spending their full time in educational and scientific institutions. According to a conservative calculation (which excludes industrial research and training establishments as well as all personnel not involved in teaching, research or learning), the proportion was 6.7 percent, or nearly 12 million people, in the European Community in 1971. If the corresponding US figure of 11.4 percent indicates a trend, the development is by no means finished.[6] This is of course a highly heterogeneous category, including professors and teachers, research directors and research assistants, undergraduates and graduates. But the point remains that the institutional order of education and science has acquired a considerable social significance in quantitative as well as qualitative respects.

The causes of this significance are those which explain the volume of scientific expenditure: an (often implicitly) assumed relationship between extensive economic growth and educational expansion. But whereas the change of subjects of research and its applications which I suggested earlier cannot be effected from within science alone, but needs political decisions, the educational class has been instrumental in bringing about a subtle change of the general social mood. Whether it is the relative, and often privileged seclusion of institutions of education and scholarship, or the notoriously anti-cyclical approach to patterns of established power by intellectuals and those who aspire to this epithet, or the frustrating experience of many that the uses of education in a growth world are rather more limited than they had anticipated — in any case, those working in colleges and universities, academies and research institutes have contributed to spreading doubt about the virtues of quantitative expansion and stimulating the desire for a greater emphasis on the quality of human life.

To be sure, I am not trying to say that the educational class has deliberately and systematically espoused such views nor even that there is such a social category in any strict sense of the word 'class'. But in those societies which do not maintain a monolithic appearance of growth addiction the political community has begun to suspect the scientific and educational communities of being subversive rather than helpful. In the negative case, this leads to the isolation of science and higher education from the structures of economy and polity; and conversely to a reduction of public interest and money for the institutions of science. I have intimated that there are traces of this happening in parts of Europe. But the negative case need not prevail; societies are capable of developing their structures to respond to a new potential without going through the more extreme forms of alienation and eventually through revolutionary upheavals. One contribution of science to such gradual change is through research projects which face the issues of future responses to the challenges of survival and justice squarely; this was my own intention in launching the preparation of a major research project 'Europe Plus Thirty' which is to mobilize the scientific imagination assembled in the European Community. The contribution of politics to the gradual departure from quantitative expansion is more difficult, although one can hope that some of the problems facing political leaders today will leave no other choice to them but to embrace a new society of intensive and qualitative growth. In the meantime, the position of science and technology in the advanced societies is likely to remain precarious for some time to come. It will be a test of the maturity of societies whether they can avoid a deterioration of the uneasy relationship between science and politics into a pointless confrontation of the future and the past.

NOTES

1. Facts and figures in this paragraph were provided by the Directorate General XII of the Commission of the European Communities (document of 6 June 1974). R & D expenditure is defined by the standards of the (OECD) Manual of Frascati. CREST = Committee for Research, Science and Technology.

2. D. Bell, *The Coming of Post-Industrial Society* (New York, 1973), 408. The earlier quotations are on pp. 14, 20.

3. Cf. R. Heilbroner, *The Human Prospect* (New York, 1973).

4. Cf. J. Maddox, *The Doomsday Syndrome* (London, 1972).

5. D. Gabor, 'The Transition to a Mature Society' (unpublished ms. of a lecture at the Hague on 11 June 1974).

6. Figures calculated by the Statistical Office of the European Communities (note by Miss Fürst of 6 June 1974). The calculation is impaired by a number of difficulties of categorization and international comparison which limit its value.

4

SOCIAL SIGNIFICANCE OF THE SCIENTIFIC AND TECHNOLOGICAL REVOLUTION

Pyotr Fedoseyev
Academy of Sciences of the USSR, Moscow

The discussion of the theme 'Science and Revolution in Contemporary Societies' at the Eighth World Congress of Sociology held in Toronto, Canada provided an arena of debate on the most significant problems of our time. Of course, the answers to these basic contemporary problems for people of differing world outlooks cannot be identical. More or less generally understood is the concept of revolution meant as a radical, qualitative change in contrast to evolution as gradual growth or merely quantitative change.

Revolution is often understood as isolated, though fundamental, changes in technology, in management and in the form of the organization of production, changes in the way of life and culture, upheavals in science and philosophy. This interpretation of the word 'revolution' is quite legitimate. But the truly revolutionary character of these changes can be correctly understood only if their social significance is spelt out, only if they are brought into correlation with the profound processes of social development underlying the mounting social revolution. There are therefore no grounds for reducing the revolutionary development of the contemporary world to revolutionary processes in the sphere of science and technology, as has become the fashion among Western sociologists.

It should be noted right away that by social revolution we mean the world revolutionary process made up of the socialist, democratic and national liberation revolutions. A distinctive feature of this historical process is that the revolutionary transformations in the contemporary world are unfolding against the background of, and in inter-connection with, the spreading scientific and technological revolution, unparalleled in scale and in its results and con-sequences.

In order to study and understand the sources and manifold impact of the scientific and technological revolution on social life and the destiny of peoples, it is by no means sufficient to conduct research into the study of science itself, or to have a knowledge of the major changes in the techniques and technology of production, or even of the changes that have taken place in the socio-professional pattern of the population of the people's educational level and qualifi-cations. Taken separately, this and other knowledge of the processes that have been generated by the scientific and technological revolution stir the imagination but give no understanding of the substance of that revolution, of the new possibilities that it is creating for mankind's progress. It has become possible to ascertain the substance and historic significance of that revolution only on the basis of a com-prehensive, truly composite approach to its assessment against the background of its individual link with funda-mental social processes.

Social changes and the scientific and technological revo-lution are closely interrelated sides of an exceptionally dynamic historical process that is unfolding apace. The specific content of both the social and the scientific and technological revolution can be correctly understood only if examined in their unity. The mainsprings of social revolution are the contradictions in the development of social production — between the productive forces and the relations of production. These contradictions are becoming particularly acute and dynamic owing to the scientific and technological revolution. As for the latter, its development depends to a decisive degree on the conditions of the social and economic system, and the more intensive the progress of science and technology, and correspondingly its impact on the development of production and all the other aspects of social life, the greater this dependence.

The development of the productive forces in the conditions of the scientific and technological revolution cannot be estimated one-dimensionally. An understanding of the scale, of the quantitative indicators of production growth, and scientific discoveries, of man's power over the forces of nature does not make the question of utilization of this power superfluous; on the contrary, it emphasizes the necessity for posing this question. The unprecedentedly increased possibilities of creation and, unfortunately, of destruction, contained in the present level of technico-scientific progress and its prospects, compel not only scientists and political figures, but ordinary people as well to ponder over the question and ask with concern: whom does this power serve, and what objectives? Good or evil? Universal welfare and human progress or the selfish interests of the monopoly groups, of the military-industrial complex? Thus, the problems of the scientific and technological revolution are seen as social problems, as problems whose solution hinges on the prospects of a social revolution, on the success of the present world revolutionary process.

The Marxist conception of the historical process, the

materialist conception of history, constitutes the scientific basis for an integral examination of the development of social production and of the forms of association and activity of people. By proving that the pattern of any historically concrete society is determined by the achieved level of development of the productive forces and the system of production relations, which constitutes the form of this development, Marxism furnished the methodological prerequisites for a thorough study of social being, the ways of its transformation and, in particular, the place and role of science and technology in the social process. Drawing attention to the growing role of science in the development of production, Marx put forward the proposition that as machine production advances and is transformed into a sphere of the technological application of scientific knowledge, science becomes a direct production force.[1] Under the mechanical mode of production, he wrote:

> practical problems arise for the first time which can be resolved only scientifically. For the first time experience and observation — and the imperative needs of the process of production itself — have reached a scale which allows for and necessitates the application of science.[2]

The further development of industry and other branches of production, and especially the scientific and technological revolution which began to unfold in the 1950s, have fully confirmed this conclusion.

Marx also showed the interconnection between revolutionary changes in production and revolutionary transformations in the social structure. He pointed out that once a revolution has taken place in the productive forces, which is a technological revolution, there also takes place along with it a revolution in the relations of production; and in this connection he explained that the revolution in production relations is not the automatic outcome of the development forces, but is the outcome of the struggle of progressive classes (in contemporary history, the working class) against the obsolete social system. Engels formulated the main objective laws of development of scientific knowledge, which today find clear-cut quantitative expression. Finally, Lenin thoroughly investigated many aspects of the revolution in the natural sciences which unfolded at the turn of the twentieth century and which served as the precursor and source of the scientific and technological revolution of our time.

It is therefore not at all accidental that Marxist social thought was the first to give a theoretical interpretation of the scientific and technological revolution and of the social and economic processes connected with it. This concept itself was first put forward by Marxists in the 1950s when a big leap forward was made in scientific knowledge and its technological application, which Western sociologists called the 'second industrial revolution', meaning by this actually only the modernization, albeit a very radical one, of the technical basis of large-scale industrial production. Already in the documents of the July 1955 plenary meeting of the CC CPSU the concise term 'scientific and technological revolution' was used to define the whole complex of new phenomena and processes connected with the rapid progress of science and technology and its results.[3] A comprehensive and clear-cut characterization of the content and perspective of the scientific and technological revolution in the conditions of two opposite social and economic systems was also given in the programme of the CPSU, adopted by its 22nd Congress in 1961.[4] Thus, it was the Marxists who from the very inception of the scientific and technological revolution were able to determine its essence and social role, and to foresee its general impact on social development.

The scientific and technological revolution is basically the radical qualitative reorganization of the productive forces as a result of the transformation of science into a key factor in the development of social production. Increasingly eliminating manual labour by utilizing the forces of nature in technology, and replacing man's direct participation in the production process by the functioning of his materialized knowledge, the scientific and technological revolution radically changes the entire structure and components of the productive forces, the conditions, nature and content of labour. While embodying the growing integration of science, technology and production, the scientific and technological revolution at the same time influences all aspects of life in present-day society, including industrial management, education, everyday life, culture, the psychology of people, the relationship between nature and society.

The Marxist methodology in analysing this phenomenon is characterized by a comprehensive, integrated and systems approach. In order to understand the scientific and technological revolution, its roots and many-sided effect on the life of society and on the destinies of people, science policy studies alone are not enough; nor for that matter is knowledge of the major changes in technology and the production processes and even of the changes in the social and occupational pattern of the population, in their education and qualifications. A knowledge of these and other processes generated by the scientific and technological revolution taken separately do not give an understanding of its essence, do not enable one to assess the prospects for its future development, and the new possibilities for human progress created by it. Only through a comprehensive, truly synthetic approach to appraisal of the scientific and technological revolution in close relationship with basic social processes can its essence and historical significance be adequately determined.

The deep-going social consequences of the scientific and technological revolution are due not so much to the advance of science and technology as to socio-economic factors. This revolution is a world-wide phenomenon, but its forms and social consequences differ fundamentally in different social systems. The difference in the content and significance of the processes of the scientific and technological revolutions in the two opposite socio-economic systems is predetermined by the fundamental differences in the goals of socialist and capitalist production. It is the position of the working people in production and in the system of its relations, in the first place, that most sharply reveals the basic differences between the processes of the scientific and technological revolution in differing social systems.

In capitalist production, Marx remarked, 'the producer is . . . controlled by the product, the subject, labour which is being embodied by labour embodied in an object, etc.'[5] This is the source of theories exaggerating the role of science and technology and endowing them with a life of their own and with power — beneficial or demonic — over the destinies of society and people. These mythological constructions suffer from the same flaws that Marx revealed in bourgeois political economy.

> In all these conceptions [he wrote] past labour appears not merely as an objective factor of living labour, subsumed by it, but vice versa; not as an element of the power of living labour, but as a power over this labour. The economists ascribe a false importance to the material factors of labour compared with labour itself in order to have also a *technological* justification for the *specific social form*, i.e., the *capitalist* form, in which the relationship of labour to the conditions of labour is turned upside-down, so that it is not the worker who makes use of the conditions of labour, but the conditions of labour which make use of the worker.[6]

The current bourgeois conceptions of the scientific and technological revolution, for all their differences, actually rest on the same methodological foundation. Depreciation of the role of the masses in the historical process in general, and of the working people as the 'human element' in production, in particular, are characteristic of both the 'optimistic' (representing scientific and technological progress as a universal means for overcoming all contradictions and curing the ulcers of bourgeois society), and the 'pessimistic' (foretelling the inevitable doom of mankind on the way to 'technological civilization') variants of the bourgeois interpretations of the scientific and technological revolution. A common feature of these conceptions is the exaggeration and absolution of the significance of science and technology in social development; these are seen as an absolutely independent and decisive factor of contemporary history, standing above society.

Philosophico-sociological conceptions of this kind sometimes appear quite convincing. Two circumstances contribute to this: first, the fact that some scientifically established data on the phenomena and tendencies in the technico-scientific process in the conditions of capitalist production are included in the pattern of such conceptions, thereby creating the deceptive impression of their being scientifically substantiated on the whole; secondly, the fact that these conceptions are tailored to the maximum degree to the fears and hopes — largely illusory — widespread in bourgeois society, induced by the rapid progress of science and technology and its obvious consequences which affect also the direct interests of the individual.

The exaggeration of the role and place of science and technology in social development, representing them as a force alien to the interests of most people or, at any rate, operating independently of their will and wishes, reflects

what may be called the sway of monopoly capital, which in
the capitalist countries has usurped the national wealth and
the means of its production, and which uses their unlimited
power in its own interests and to the detriment of the
interests of the labouring majority. Furthermore,
absolutization of the role of science and technology in social
development reflects the scepticism of bourgeois sociologists
towards the ability of society to take under its control the
spontaneous development of technicalized capitalist production. The advocacy of such conceptions has as its purpose
to induce the mood that man is powerless in a 'technicalized
society', that social movements, the struggle for social
progress and the abolition of exploitation and alienation have
no future. Finally, many of these conceptions perform an
apologetic function, that of seeking to remove from
capitalism the blame for the misfortunes and sufferings of the
working people, for the injustices and vices of this social
system, for the mass destruction of human potentialities and
the very habitat of man, and to hold science and technology
responsible for all this.

The apologists of the bourgeoisie have over the years been
stubbornly inculcating this idea which is retrogressive in its
very essence. As far back as 1927 Einstein came out against
it. In his article 'Isaac Newton' he wrote that the intellectual
tools, without which it would have been impossible to
develop modern technology, came mainly from observation
of the stars, that for the misuse of this technology the
creative minds, like Newton's, were no more responsible
than the stars themselves whose contemplation inspired their
thoughts. Science and technology are created by the
efforts of people, and it is precisely upon them, upon the
social organization of their activity that the manner in
which scientific and technological achievements are used
depends. And science and technology for their part, although

they reflect particularly graphically the way the productive forces develop, do not function without people; they act as productive forces only to the extent that they join with living labour, only in activity of the working people, organized and directed in conformity with the objectives of a given society and the mode of production prevailing in it.

We proceed from the fact that people, with their experience and know-how, with the historically determined forms of their labour co-operation constitute the principal productive force, while technology is materialized labour. Under capitalism, this materialized labour is alienated from the producers and becomes a means for alienating living labour. Under socialism, where the aim of production is totally different, man, retaining and increasing his importance as the main productive force, ceases to be a means of production and an appendage of technology.

It is essential strictly to observe the difference between the concepts 'the productive forces' and 'technology'. Taking issue with ideas of this kind, scientifically refuting absolutization of the significance of scientific and technological progress, or exaggeration of the role of technology as very nearly the only important component of the productive forces, is nothing new to Marxism. In *Capital* Marx stressed the need to distinguish such different things as the imminent features of machine production and the capitalist application of machines.[7] Science and technology must be an object of theoretical analysis only in the context of the whole of social development, while scientific and technological progress and its consequences can be correctly understood and interpreted only if an integral analysis is made of society's material foundation — its productive forces and production relations taken in the aggregate, in their interconnection.

From about the last third of the nineteenth and the early twentieth century the development of the productive forces had been leading at an accelerated rate to the socialization of production, to individual private enterprises being ousted and replaced by big industrial, trading and financial corporations and the state economic sector. While on the technological plane the growth of the productive forces necessitated that essentially technical production functions be increasingly transferred from man to machines, on the socio-economic plane this growth demanded the socialization of production. This, in turn, created the conditions for the unprecedentedly rapid development of such a component of the productive forces as science, since at the monopolistic stage of capitalist development, as Lenin stressed, 'the process of technical invention and improvement becomes socialised'.[8] The high degree of concentration and centralization of the economy determined the growing scale on which science is applied in material production and, subsequently, what may be called the industrialization of research and development. Ever since the development of industry, the concentration and centralization of production began to be effected through the increasingly broader application of machinery, the introduction of new technology in material production has served as the principal instrument for developing the productive forces. At the same time technology has increasingly become not merely materialized human labour, but materialized *scientific* knowledge. All this has radically changed the relationship between science, technology and production and has led to the creation of giant scientific-technological and industrial complexes.

   If we examine on this level the evolution of the relationship between science and technology we will find that the introduction of scientific discoveries in industry has proceeded at an unprecedented rate. When isolated private enterprises prevailed, when scientific quests were the concern of individuals lacking the necessary production facilities, it took decades to apply scientific discoveries on a large scale. As the industrial enterprises forming the base for the technological application of science were amalgamated and centralized, scientific discoveries began to be applied in industry more and more quickly. This may be seen from the following data showing the narrowing of the gap between a scientific discovery and its application: in the case of photography — 112 years (1727-1839); electric motor — 57 years (1829-86); telephone — 56 years (1820-76); radio — 35 years (1867-1902); electronic valve — 31 years (1884-1915); X-ray tube — 18 years (1895-1913); radar — 15 years (1925-40); television — 12 years (1922-34); nuclear reactor — 10 years (1932-42); transistor — 5 years (1948-53); and solar battery — 2 years (1953-55).

The acceleration of scientific and technological progress is even more evident in the growing connection between the appearance of new fundamental scientific discoveries and the emergence of such major branches of modern industry as the peaceful use of atomic energy, the development of computers, and space exploration.

Thus, the growth of socialization and of the scale of production, which has made it possible to make big investments in science and to build powerful research-industrial complexes, has served as the economic basis for the scientific and technological revolution. The development of the productive forces, and together with it of the socialization of production, are the material prerequisites for scientific and technological progress as well as for fundamental social change.

It is not accidental that the social revolution and the scientific and technical discoveries of our century coincide in time, for they are induced by the growth of the productive forces and the need for their further development. Capitalism, relying on the availability of a developed industrial base and skilled cadres, has created tremendous productive forces, which have reached a high degree of socialization and concentration, including the establishment of international corporations. Yet capitalism cannot solve the basic social problems, especially the problem of social inequality and the division of society into the 'haves' and 'have-nots'. It is precisely the process of the socialization of the productive forces, which reaches its highest stages under capitalist machine production, that lays bare the main contradiction of private-capitalist society — the contradiction between the productive forces and production relations. That is why large-scale industry which develops under capitalism makes essential, to quote Engels,

> the establishment of an entirely new organisation of society under which the management of industrial production is effected not by individual manufacturers who are competing with each other, but by the whole of society in accordance with a definite plan and with the needs of all members of society.[9]

The scientific and technological revolution, clashing with the obsolete system of social relations, has further aggravated the contradictions of capitalism. Lenin noted with keen insight that under monopoly capitalism 'the extremely rapid rate of technical progress gives rise to increasing elements of disparity between the various spheres of the national economy, to anarchy and crises'.[10]

While it engendered an unprecedented growth of the productive forces, capitalism, at the same time, caused the unprecedented sharpening of the social antagonisms inherent in the private capitalist organization of production. Besides the basic antagonism between the social character of production and the private-capitalist form of appropriation, there is the deepening conflict between the commercial interests of the monopolies and the habitat of man, between the rapidly developing technology and nature, between the vital requirements of people and the unchecked growth of the means of destruction which threaten the very existence of mankind and life on earth.

The scientific and technological revolution, far from removing the need for social revolution, makes such a revolution even more imperative. Only revolutionary changes in society make it possible to abolish the obsolete social relations and establish new forms of industrial management, characteristic of socialism, by socializing the means of production.

Bourgeois ideologists cannot deny the revolutionary character of our epoch, but they are endeavouring to reduce the modern world's revolutionary development to scientific and technological progress, to changes in science and technology, ignoring the revolutionary processes in socio-economic relations or seeing the direct cause of social changes in the self-development of technology. It is no accident that in bourgeois social science the traditional idealistic concepts are giving way to so-called technical determinism, according to which the fundamental cause of social changes lies in technological advancement.

A typical feature of the latest non-Marxist theories is that they divorce scientific and technological progress from the concrete social conditions of that progress. These theories belittle the role of the people in historical development generally, and the role of the working people in industrial progress in particular. They bury in oblivion the obvious truth that machinery is created by people and is the motor of production only when it is operated or controlled by man. Although machinery is important and plays an ever-growing role, people who create this machinery and who realize its possibilities remain and will always be the principal productive force.

When the people's active, creative role is disregarded machinery is counterposed to man or, conversely, man is counterposed to machinery. While some ideologists regard machinery as the panacea for all ills, declaring that it can do everything for man, others, on the contrary, believe that machinery operates against man and consider that it is the cause of all calamities, that it is a satanic force hostile to man and urge that it would be to man's good to renounce technological progress. This attitude expresses a lack of faith in man's ability to direct the development of science and technology rationally. The aggravation of the crisis phenomena and contradictions in bourgeois society as a result of scientific and technological progress are due, according to these ideologists, not to the vices of the capitalist system but to the increase in machinery. They oppose the Marxist teaching on the uncompromising struggle between the exploiting and exploited classes with pseudo-scientific arguments about a conflict between man and technical progress. At the last International Philosphical Congress, a philospher asked: 'Have we not come to the replacement of the Marxist concept of exploitation of man by man with a technicized form of man's exploitation by machines?'

But the scientific and technological revolution is not the result of an artificial attraction to machinery. It is the result of the powerful development of the productive forces and the growing concentration and centralization of production. Under capitalism these processes objectively lay the ground for the transition to the higher, socialist system. However, the bourgeois economic system usually deforms scientific and technological progress and, at the same time, is an obstacle to social advancement, as a result of which all the contradictions and vices of that system grow more pronounced.

In bourgeois society the principal contradiction of the scientific and technological revolution is that machinery is separated from man by capital. It is the property of a handful of owners of means of production and serves as the means for exploiting the working masses. The historical way out of this situation lies not in renouncing technological progress or in hoping for the automatic removal of all the social evils springing from technological development. The solution lies in turning the means of production into public property in the interests of the working people, of the whole nation.

The conclusions drawn by Marxist-Leninist theory from a generalization of historical experience show that scientific and technological achievements provide the objective condition for the development of production, thereby facilitating the creation of the material conditions for social advancement. But in themselves they cannot cause historical changes or determine the course of world development. The masses headed by the international working class, whose cardinal gain and mainstay is the socialist world system, are the principal motor of social progress, of all far-reaching revolutionary changes.

On this point Marx said in his speech at the anniversary of *People's Paper* (14 April 1856): 'We know that to work well the new-fangled forces of society, they only want to be mastered by new-fangled men — and such are the working men.'[1] The victory of socialism in the USSR and the socialist transformations in a number of other countries comprise an epochal achievement of the working class, which heads society's progressive forces.

Moreover, the masses are the principal force of national

liberation. Awakened by the Great October Socialist Revolution, the historic achievements of socialism in the USSR and the victorious outcome of the war of liberation against German fascism and Japanese militarism, the peoples of colonial and semi-colonial countries entered the world scene as an independent creative factor and struck a crushing blow at the imperialist system of international exploitation and oppression.

The masses are the main force defending democracy and human rights. Together with all strata of the working people, the working class upholds democratic liberties against encroachment by the reactionaries, closes the road to an undisguised terrorist dictatorship of capital, and selflessly combats neo-fascism and reactionary fascist regimes.

Led by the working class the masses are the main social force in the struggle for peace and international security, against the threat of another world war, against all forms of aggression and local wars. The present world peace movement was brought to life under the motto: 'The defence of peace is the cause of all the peoples of the world.' The more actively the peoples defend peace, the greater will be the guarantee that another war will not break out. As we understand it, peace is not only a question of security but also a prime condition for the solution of major present-day problems affecting mankind's very future: the problem of energy sources, environmental protection, the eradication of evils such as mass starvation and dangerous diseases, and the tapping of the ocean's wealth. Mankind's problems of the future and many urgent present-day tasks cannot be resolved outside the system of international relations founded on peaceful coexistence.

From this springs the fundamental significance of the conclusion that the cause of peace, democracy, social progress and public well-being and the prospects for mankind's further development ultimately depend on the mass of working people, on their consciousness and activity. It would be a fatal mistake to underestimate the role of scientific and technological progress in modern world development. The Communists emphatically reject the primitive prejudices of the Leftist doctrinaires against the scientific and technological revolution. A correct understanding of its role and effects is of immense significance to theory and practice. This is particularly important today when, in connection with the scientific and technological revolution, the future of all mankind is broadly prognosticated, when scientific and technological progress is the principal lever for the creation of the material and technical basis of communism, when this sphere of progress embraces one of the principal sectors of the historic competition between the two opposing social systems — between socialism and capitalism.

The socialist system of economy ensures the purposeful application of scientific and technological achievements and the systematic improvement of social relations. Under the socialist system scientific and technological progress combines harmoniously with social progress, thus providing the opportunity to achieve social equality, to humanize labour and develop the capabilities and talents of the individual as fully as possible.

With that end in view, the 24th Congress of the CPSU set the task of combining the achievements of the scientific and technological revolution with the advantages of the socialist economic system.

The scientific and technological revolution, a powerful motive force, creates the possibility for universal prosperity, for a better social structure, for improved living and working conditions, for the all-round development of the personality and for protecting the natural environment. The success of material production creates the conditions for expanding the sphere of spiritual culture, for new forms of leisure and for the intellectual development of every member of society. However, all these new possibilities are not realized automatically. If the scientific and technological revolution is not directed and regulated, it may give rise to serious imbalances and contradictions in social development. Therefore, a major task of mankind today is to learn how to tackle, on a planned and purposeful basis, the new economic and social problems constantly posed by the scientific and technological revolution.

The objectives of the social development of society is a crucial issue. First in importance, we believe, is to provide the conditions for social equality and for society's steady advance towards homogeneity. Here, it should be emphasized that by social equality we mean the same conditions of labour and consumption for all, and not the artificial levelling of the capabilities, tastes, requirements and interests of all people. It is a question of closing the gap between poverty and wealth, between ruling and oppressed classes, and of eliminating the distinctions in the education, living conditions and way of life of manual and mental workers. Scientific and technological achievements create the material prerequisites for resolving these problems. But by themselves they cannot change the class structure of society; only in combination with a social revolution can scientific and technological progress achieve real social equality. Marxism also links the problem of the development of the individual and his emancipation with the abolition of society's division into opposed classes and with the obliteration of class distinctions.

By establishing public ownership, socialism has for the first time made it possible to abolish exploiting classes, to improve the social structure and social relations so as to

achieve complete social equality and the all-round development of the individual. Of vital importance for narrowing
the disparities in the living standards and way of life of the
different sections of the population under socialism is the
growth of the social consumption funds which allow for
gradually decreasing the elements of material inequality
still existing under the socialist principle of distribution
according to labour. In the USSR, per capita payments
and allowances from the social consumption funds increased
from 127 rubles in 1960 to 310 rubles in 1973. This money
is spent on free education, medical service, housing, and on
providing good living conditions for the categories of the
population not engaged in labour (pensioners, invalids,
students). In the last three years alone increased pensions,
scholarships and other allowances have meant increased
incomes for 23 million people.

Under socialism the scientific and technological revolution
opens up new possibilities for promoting the individual's
creative abilities, for intellectualizing labour and humanizing
living conditions. This is facilitated, in particular, by the state
system of all types of training and education which are free
and accessible for everyone in the socialist countries. At
present, three-quarters of the population engaged in the
national economy of the USSR have a secondary and higher
(complete and incomplete) education. In 1973, 3.7 million
boys and girls received a secondary education, that is, nearly
three times as many as in 1963. The country's higher
educational institutions are attended by 4.7 million students,
and its specialized secondary schools by 4.4 million students,
that is, twice as many as in 1960. With the appearance of new
professions an extensive network of advanced training has
been launched in the country. In 1973 alone more than 20
million people acquired new professions or improved their
skills.

Socialism ensures the broadest masses access to education. In particular it gives the children of all social strata equal opportunities to receive a higher education. Western sociologists have calculated that in capitalist countries the children of the bourgeoisie have ten times more chances of entering university than their coevals from lower social strata. In the USSR and other socialist countries the reverse is the case; it is the children of working people who enter universities and institutes, for there are no longer any exploiter and parasitic strata. Young workers and collective farmers who have been working in production are given priority rights when entering university. Besides, free one-year preparatory courses are open to them which give them the necessary grounding for entering a higher educational institution.

The scientific and technological revolution poses new problems also as regards the conception and development of democracy. Application of the latest achievements of science and technology in management calls for a high level of specialized training and, consequently, for ever greater professionalization. In bourgeois society this leads to the intensification of technocratic tendencies, to increasing alienation of the system and apparatus of management from the people, from parliamentary and generally representative bodies of democracy, to the actual concentration of all power in the hands of executive bodies. In the Soviet Union the professionalization of the executive bodies of management is combined with enhancing the role of the representative bodies — the Soviets at all levels and the social and political mass organzations: party, trade union and youth organizations, production conferences, and so on. In the socialist countries systematic work is conducted to develop the activity and effectiveness of all democratic bodies in every possible way, to ensure their control over the activity of the executive organs of government and to draw, on an increasingly larger scale, the masses into the administration of state affairs. This is facilitated by raising the general and the political education of the population, their cultural and professional level, by their growing social consciousness and maturity.

Today the world is becoming more and more convinced that many of the negative tendencies observed in the industrially developed capitalist countries, such as the growing exploitation and intensification of labour, the monotony of labour processes, chaotic and unplanned urbanization, depletion of natural resources and pollution of the environment — that all these tendencies are not the fatal and unavoidable outcome of technological progress as such, but the consequences of a social system that gives priority to technical and economic growth for the sake of maximum profits, regardless of the price people have to pay.

Western sociologists analysing the social consequences of the scientific and technological revolution offer no clear-cut programme for utilizing that revolution in the interests of man. In their forecasts they often draw attention to the adverse consequences produced by the application of technology and warn against imminent critical situations, but fail to show how and on what economic basis the social processes taking place during the scientific and technological revolution can be controlled. Advocates of the technocratic and scientistic trend, which is widespread in the West, hold that scientific progress in itself, involving no social changes, is a panacea for all ills that automatically solves all the problems facing mankind. Not so the advocates of the opposite, the anti-scientistic trend, who take the stand of abstract humanism. They refuse to recognize the great new possibilities opened up by the scientific and technological revolution and blame it for all the social ills. The metaphysical alienation of scientific and technological progress from its concrete historical conditions leads to the imaginary dilemma: 'Either the progress of science and technology or the progress of man.' The views of the proponents of these two different trends — whether they are enthusiastic or pessimistic about scientific and technological progress, or even if they are dead set against it — reflect their helplessness in resolving the new social problems produced and aggravated by the rapid changes in modern science and technology.

All these problems cannot be reduced to any one aspect — man and nature, man and technology, rationalization of life and freedom of the individual, and so on. They call for, we repeat, a comprehensive, systems approach because the scale and rate of the changes brought about by the scientific and technological revolution imperatively necessitate foreseeing in good time and as fully as possible the overall consequences of scientific and technological progress in the sphere of production, the economy, as well as in the social sphere and its impact on society, nature and man himself. In our approach to this problem we do not share the view of the distinguished English historian and philosopher Arnold Toynbee who holds that technological progress has never been accompanied by corresponding moral and intellectual progress, that morals are static, while technology is dynamic, and that the gap between the technical level and the social and moral advance of mankind is constantly widening. In tackling the problems of contemporary civilization we proceed from the idea of the unity of scientific, technological, social and moral progress.

The scientific and technological revolution increases the social and moral responsibility of people since the forces they have created can threaten their own existence. Today the problem of freedom is not just a matter of personal choice and of responsibility for it; it is a problem of men's responsibility in determining the paths of development of human civilization, which is capable of ensuring not simply the survival, but also optimal living conditions for the future generations.

It is relevant to recall that Marx scrutinized these problems more than a hundred years ago, favourably evaluated the book by Karl Fraas, *Climate and the Vegetable World Throughout History. A History of Both,* which proved that in historical times climate and flora have changed under the influence of man. Marx noted the unconsciously socialist tendency of the conclusion drawn by the author that when culture develops spontaneously it leaves a desert behind.[12] But he also noted the bourgeois narrow-mindedness of Fraas in failing to see that such devastation could be avoided if the development of culture was consciously directed by society.

The narrow-mindedness Marx discovered in the criticism of the destructive tendencies of scientific and technological progress is evident also today in the way some Western scholars interpret many human problems stemming from the material and technical achievements of mankind.

This narrow-mindedness can be got rid of and the dialectical interconnection of the productive forces and the system of social relations can be understood only by proceeding from an integral perception of the world and one oriented towards the revolutionary reshaping of the world in conformity with scientifically substantiated humanistic ideals. Herein lies the key to understanding the objective laws of the social and scientific-technological revolution of our time, to understanding their interconnection and interaction. In the conclusion of my report I cannot help expressing my deep sorrow that in some reports included in this volume, namely by Professor Touraine, there are some political statements not connected with science and falsifying the essence of international and inner policy of the Soviet Union.

NOTES

The study, 'Social Significance of the Scientific and Technological Revolution' submitted by Pyotr N. Fedoseyev was presented in Toronto by Mikhail N. Rutkevich.

1. See K. Marx, *Capital*, Vol. III (Moscow, 1971), 81.
2. K. Marx and F. Engels, *Works*, 47, 554 (in Russian).
3. See *Pravda*, 17 July (1955).
4. See *Programme of the Communist Party of the Soviet Union* (Moscow, 1961), 26.
5. K. Marx, *Capital*, Vol. IV, Part III (Moscow, 1971), 275.
6. Ibid., 275-76.
7. See K. Marx, *Capital*, Vol. I (Moscow, 1971), 415.
8. V.I. Lenin, *Collected Works*, Vol. 22, (Moscow), 205.
9. K. Marx and F. Engels, *Works*, 4, 329 (in Russian).
10. V.I. Lenin, *Collected Works*, Vol. 22, (Moscow), 209.
11. K. Marx and F. Engels, *Selected Works*, Vol. I (Moscow), 501.
12. See K. Marx and F. Engels, *Selected Correspondence* (Moscow, 1955), 243.

5

SCIENCE, INTELLECTUALS AND POLITICS

Alain Touraine
C.N.R.S., Paris

INTRODUCTION

The general theme of the Eighth World Congress of Sociology, 'Science and Revolution in Contemporary Societies', is fortunately ambiguous, because it obliges us to side with one or the other of its possible interpretations. Must one understand it, as is suggested by the title of the inaugural session, in the following manner: Mankind has experienced an unprecedented scientific and technical revolution, a fundamental modification of its forces of production. How will the social organization adapt itself to this transformation? What new relations of production must appear? What type of crisis will be experienced by those forms of society rendered archaic and irrational by the advance of science and technique? Or must one, on the contrary, suppose that revolutionary movements are not brought about by scientific and technical progress, but rather exert themselves against a domination which relies on technological power and therefore on scientific development? At first, one is tempted to identify this opposition of points of view with that of antagonistic social regimes. The theme of the scientific and technical revolution and its social consequences is undoubtedly at the center of the reflections of many sociologists from socialist countries, as has been shown by Radovan Richta's statement. It is also true that the theme of revolutionary forces was included in the general theme of the Congress — yet separately from the progress of science and technique — by sociologists from capitalist countries and from countries dependent on the international capitalist market.

However, this identification of the two possible inter-pretations with the opposition of socialism and capitalism is more tempting than it is solid. One has difficulty imagining that a sociology of Marxist origin would reduce an analysis of revolutionary forces to the progress of science and tech-nique; moreover, many elements show that capitalist countries have experienced, in the last two decades, a great confidence in the scientific and technical revolution, and have often accepted the idea that societies should 'adapt' themselves to the new forces of production. These themes have been widely diffused by the ruling classes and their ideological mainstays. Personally, I am conscious of the fact that post-industrial society was conceived within this techno-cratic spirit; because I have fought this conception and tried to define post-industrial society as the site of new powers and conflicts, I realize to what extent my critical position is a minority position, enveloped by an optimism which announces the 'end of ideology'.

We find ourselves, therefore, before two conceptions of society. According to one of them, scientific and tech-nological rationality imposes, more strongly than ever before, the creation of a rationally organized and planned society, rid of all irrational and 'particularist' forms of power. According to the other, science and technology are the new power base for the great economic, political, and military apparatuses, which impose their technocratic interests in the name of that rationality which they have appropriated; and which reject or confine those who fight their power. This means that one should observe the formation of new revolutionary movements, which question the identification of science with apparatuses of domination, which oppose a certain conception of science, and which want to obtain a collective reappropriation of the new forces of production by newly expropriating the expropriators.

Each one of us is placed before this choice which is, at one and the same time, an intellectual production and a political decision. This means that sociologists must decide whether the idea of society must definitely disappear, whether society should be transparent to scientific and technical forces — since the study of social organization is, in this perspective, nothing more than the study of obstacles and delays related to the demands of rationality — or whether on the contrary, the scientific and technical forces of production should appear, much more completely than the antecedent ones, as what is at stake in social relations, decisions and conflicts. Is technique dominated more than ever by politics or is politics becoming the apex of technical rationality?

THE SOCIOLOGY OF THE GODS

All societies at least from the moment at which they have acquired important means with which to act on themselves to transform their performance, have conceived their existence as determined by a superior order, a meta-social order, be it a divine will, principles of political philosophy or economic 'laws'. The distance between social practice and these meta-social guarantors of the social order diminishes unceasingly as society's capacity to transform itself increases. On the other hand, these meta-social orders were not separable from their opposite, that is, from the acknowledgement of an ascribed, even a natural, social and cultural situation. Social action was wedged between the meta-social order and the natural order. The world of the gods, which commands that of society, was also the world of the community and of its laws of reproduction and inte- gration. This 'social nature' deteriorated along with the meta- social order as society began to appear rather as the product of its own action. Therefore, industrialized societies no longer experience the traditional alliance between the gods and nature, yet are attracted by two opposing images of themselves that of natural society and that of rational society.

I understand this second term as the image of a society which deifies its own capacity to operate on itself. Such a society no longer believes that social facts are determined by meta-social orders; but it does believe that they are determined by science, indeed a human creation, and yet an order independent of human action; a natural order as well, yet in a non-cultural sense, since natural orders maintain themselves without the interference of values, norms, institutions, in other words, of socially regulated behavior. By natural society, one must understand the image of a society reintegrated into the natural system, and which thus subordinates its capacity to change and to produce itself to the demands of equilibrium.

If we try to situate our society in relation to others which have preceded it, we see that the field of social action has spread, driving the gods and nature further and further back, in the name of a 'civilization' built on labor and the exploitation of natural resources, even to the point of considering society as nothing more than a tool, an operation, a mere machine. It is fortunate that this productivistic optimism has been shaken; that the naïve evolutionism inherited from the nineteenth century has been rejected; that our enriched societies cease identifying themselves with universalism, rationality, and progress. But it is the opposite menace that should be discussed here, because 'progress' has only apparently caused the limits of social action to retreat and, today as yesterday, the field of social action runs a great risk of being invaded by the gods and by nature.

The power of science and technology could lead to the creation of new gods, more demanding than the old ones because more anthropomorphic. Society would be nothing more than the setting up of social and cultural conditions for the full development of scientific and technological forces of production. Social facts, instead of being analysed by themselves as a system of social relations, would have to be judged in reference to a non-social order, the scientific and technical revolution. This would inevitably produce a counter-utopia, which judges social behavior only by its conformity to the demands of the natural order, to the needs of an eco-system in which mankind would accept nothing more than a 'niche'. Here we are, once again, yet in a manner probably more extreme than in the past, wedged between the limits of social action. These seem to disappear beneath the confrontation of productivism and naturalism. We were tempted, only recently, by the image of a triumphant society, and here we are abruptly placed back into the narrow slit between the meta-social and the infra-social. At the same time, sociology is questioned as to its very existence. The pressure of socialism, in return, entails a relish for societies 'without history', concerned above all with their equilibrium and their survival.

There is great danger in referring to the scientific and technical revolution in terms which destroy the very possibility of knowledge concerning society, and which subordinate social action to laws, which are said to be scientific and yet which are nothing more than the laws of an absolute power. On the contrary, a reflection on the new forces of production could and should lead to the rejection of this technocratic interpretation and to a clearer realization that setting the forces of production into action depends on social and political conditions. One realizes then that, instead of speaking of the scientific and political revolution, one must speak of development policies.

Let us consider for still another moment the image of a society which is associated with the subordination of social action to a meta-social order, with a world of gods, whether they be the gods of religion, politics, economics, or science. This image is dominated by dualism. Social action is practised within the order of social relations, but it is commanded by a superior order. Between these two levels exists a contradiction, but there also exists 'irrational' forces of intervention, which assure the superior order of its hold over the social order. This illustration is applicable to sociologies of value as well as to sociologies of domination. Max Weber analyses the appearance and the progress of instrumental rationality in Western capitalist society. But above all, he also recognizes that action is commanded by values which put them into motion. Is not the progress of capitalist rationality provoked by religious fundamentalism? Is not political action, as calculating as it may be, oriented by an ethic of conviction?

In a similar manner, does not the presentation of society as the setting up of a social, political, and ideological class domination, suppose a constant relation to an order more fundamental than the forces of production, productive labor, and the natural needs of men? This time, it is not charisma that is the lightning bringing heaven's fire to earth; it is rather revolution, which plays an analogous role in this 'critical' version of the social order.

This dualism leads one to consider social problems as the effect of a deviation, a rupture with the superior order. A rupture created by private interests or by routine and bureaucratization. Therefore, solving a social problem always means removing social obstacles and liberating creative energy or prophetic invention. It is in this way that 'Leftist' sociology criticizes educational inequality, hoping that a greater equality will tear down the barriers which separate too many children and adults from liberation through science and learning.

In a manner at the same time contrary and parallel, the 'Rightist' critique hopes that school rules will not destroy the creative power nourished by traditions and by specific cultural and social features. In all cases, the idea is to change the institutions and adapt them to that which commands them naturally, be this Reason or Nature or a combination of the two. If science is considered the god of industrialized society, in the same way that history was the god of industrialization, and the Prince of market societies, then science is thus placed outside the field of social action and political interests. An analysis must thus limit itself to a criticism of those forms of social organization which do not conform to the demands of science and technique. But what does this mean: conform to the demands of . . . ? The only useful criterion is that of the actual development of the forces of production.

Such was in fact the Soviet leaders' idea : we will overtake and exceed other countries in the domain of production, thus proving the superiority of our regime. Such reasoning does not differ from the reasoning which, in a more diffuse manner, characterized the ideologists of the end of ideology. For them as well, it was a question of exalting a regime, which would be effective, more capable of directing and producing change, more 'secular', freer from absolutes, principles and rigidities, which are the obstacles to pragmatic adaptation. In both cases, the good society is that society which is most transparent to the demands of movement. In both cases, we are dealing with a depoliticized definition of society.

But also, in both cases, we are dealing with an entirely ideological analysis, which is at the service of the power bearers and whose principal objective, as in the case of all ruling class ideologies, is to testify to the existence of a ruling class, class relations, and domination mechanisms.

SCIENCE AND THE APPARATUS

When, in the middle of the nineteenth century, Renan expresses the bourgeoisie's confidence in the future of science, he considers science as the agent of progress, of the enlightenment. His confidence in Reason feeds his critique of the religious spirit and the ecclesiastical apparatus, at the same time that it comforts the new ruling class, the new establishment, bearers of the modernizing and humanitarian inspiration. Science was then a lay divinity, above all at the time that the book was written (1859), a long time before it was published (1890), but science was not yet a force of production. When we speak today of the scientific and technical revolution, we can no longer accept this isolation of science, a star in the firmament. What we are speaking of is the block, constituted by science, technology, and the apparatuses of production, of pure and applied knowledge.

To say that science has become a force of production is not sufficiently explicit. One must add that science, in our society, is inseparable from the apparatuses of production, and therefore from the relations of production. In an agrarian society, one cannot speak of land separately from the regime of landed property. In the same way, in industrialized societies, science has descended from heaven to earth. It is certainly wrong to say that science does not exist outside the conflict of social interests, as if there were such a thing as a bourgeois science and a proletarian science, opposed to each other even in their very principles. But it is not sufficient to reject this error and its alarming consequences; one must also distrust a naïve confidence in science, detached from private interests. Such is a professor's utopia, of which Talcott Parsons was one of the most extreme representatives, and which became the banner of the professor's middle of the road policy during the student struggles in the United States: rejection of the trustees and the pressures exerted by them; a parallel rejection of the student movement; a glorification of the world of academic representatives of science and reason. This is a false debate which hides the real problem. Science is at the service neither of the bourgeoisie nor of the proletariat, nor is it completely outside the field of social conflict; because it is incarnated in apparatuses, it has become in itself an agent of social domination or, at the very least, it is what is directly at stake in social struggles. We will try little by little to define and distinguish better these two expressions. But first the role of science as a force of production must be explained.

Any type of society must be defined by its mode of intervention in its economic activities, and by its type of accumulation and investment. For a long time, society was capable of operating only on its consumable goods, and of accumulating means for the reproduction and simple growth of produced quantities. At a higher degree of intervention, there have been merchants accumulating tradeable goods, thus intervening in the distribution of goods, activities, and incomes.

What we refer to as industrial society is that society in which capital intervenes to transform the organization of labor. Finally, there are some societies which intervene not only in its organization, but also in production itself by creating new products, by discovering methods for the management of economic organizations and communication networks, by studying the foreseeable consequences of decisions taken today, etc. Science cannot be reduced to technology. There is no direct relation between a flow of scientific inputs and a flow of technological ouputs because, between the two flows, intervenes a stock of knowledge in which scientific contributions may reside for a more or less long time before producing technological effects. But one very justly refers to the scientific and technical revolution, when one insists on the link between the two domains as a new social phenomenon. It is in the military domain that the role of science is most visible. The link btween science, technology and apparatus is more direct here than elsewhere, because contracts for research and development, a large part of which is located in the military domain, are as a general rule directly accorded to operations placed under the direct control of military and para-military organizations. Research on atomic energy, on space exploration, and on tele-communications has been and is the expression of the control of the apparatuses on scientific production itself. In more general terms, science is as much the product of a science policy as it is an independent production, whose appearance would be favored and whose discoveries would be diffused and applied by a science policy. It is dangerous to choose an extreme position in this domain. Science is not produced merely on demand, and we must remember that scholars, in order to defend their independence, stress with good reason the fact that many important techniques have emanated from knowledge elaborated independently of any kind of application program. But this distance, which always exists between the scholar and the scientific apparatus, in laboratories, universities, and hospitals, and which is essential to an understanding of the behavior of scientific workers, only seems to underline the importance of the bonds which exist between the production of knowledge and the power of the 'technostructures'.

Because the management of large organizations necessitates ever more elaborate and diverse techniques, one sometimes refers to those who produce and apply this knowledge as technocrats. This hides the principal fact. What is important is not the fact that an organization should be useful to science and technology, as if its resources were placed at the service of scientific ends, but rather that the organization be oriented by objectives, which can be symbolized by the quest for power, and which subordinate science as a means to an end.

A technocrat is neither a scholar nor a technician; he is an organization manager who tries to strengthen the power of his organization by appropriating and utilizing scientific and technical knowledge. We speak of the scientific and technical revolution and we think that we may see, at the very top of society, large organizations which place an important part of social resources at the service of science and technology.

Reality is very different and almost the contrary. The large organizations present themselves as 'pragmatic'. They constantly affirm — in a mixture of good and bad faith — that they must adapt themselves to a changing and dangerous environment. Their key word is not science, but strategy.

The development of science and technology cannot be separated from the *Realpolitik* which guides the governments and the corporations that control the production and the utilization of scientific and technological knowledge.

In those societies in which the power system is strongly integrated and centralized, the apparatus which manages the society can more easily impose its ideology, that is, its identification with the forces of production. In those societies in which power is less integrated (that is, where economic, political, and ideological power are, relatively speaking, more autonomous) and more decentralized, it is more difficult, at any one center of decision, to identify with the general interest and the reign of Reason. Therefore, each speaks rather in terms of strategy. But this difference is more important at the level of ideological behavior than at the level of social reality. In both cases, there is a real autonomy of scientific knowledge, but also and above all, the progress of knowledge is oriented according to the demands of power, whether they be those of private enterprise or those of the State.

Intellectuals perceive their role as having been profoundly transformed by this scientific and technical revolution, which should be redefined as the birth of post-industrial society, which I have more correctly named programmed society. Traditionally, intellectuals have divided themselves among four principal roles. The first consists of producing the dominant ideology, that is, giving an integrated vision of culture and society from the point of view of the ruling class. This is done in such a way that the dominant ideology escapes the field of vision and society seems to be governed by laws, principles, tendencies, and not by conflicts of interests. The second answers the first: some intellectuals contest this dominant order by revealing its character and identifying themselves with the point of view of the dominated classes. The third consists of elaborating, in a more of less autonomous manner, an interpretation of a concrete historical situation, and unifying in thought elements which would be separated by an analysis. This is what happens, for example, when one expatiates on the spirit of the Renaissance or on French culture. The final role consists of producing a model of knowledge and a cultural model with which a society may constitute its culture field, and thus the cognitive categories and values which organize its experience. This last role is never separated from the conflict between the first two; neither does it reduce itself completely to this conflict.

What is new in programmed society is that the intervention of intellectuals ceases more and more to place itself at the level of the 'super-structures', the established order, and shifts towards the level of the forces and relations of production. The interpretation of historical reality is left more and more to the mass media, which displaces intellectual-interpreters, such as Renan, whose name I cited when I began. In the second place, the intellectual-interpreter of the popular classes also tends to disappear, because the extension of the scope of politics permits the popular classes to express themselves more and more without an intermediary. Social movements at the base of society demand the capacity to express themselves directly, without having to pass through the intelligentsia. In the third place, an ever more pragmatic ruling class distrusts intellectuals, and refuses anything that seems absolute and that risks becoming a barrier to change. On the other hand, intellectuals are led to intervene more and more directly, not as ideologists but as managers. In large modern industry, much of the research has examined the position of researchers. Now, after having spoken at length of the role conflict experienced by researchers, such research has had to admit that technological production was, above all, a central element in the power of the enterprise, and that the researcher was thus ever more directly a manager, identified with the defense of the enterprise's interests (which consequently leads to strong protest attitudes, but only in a minority). Consequently, the direct reference to a cultural model, that is, to science, can no longer be separated from socially defined critical action.

The complexity of intellectual roles tends to reduce itself to a dichotomy: organic intellectuals and critical intellectuals. On one hand, intellectuals who are identified with management; on the other hand, intellectuals who associate the wish to separate national production from social domination with an affirmed solidarity with the interests and protests of dominated groups.

These observations become equally valid for natural and for social sciences, from the moment that the management of large organizations utilizes the social sciences, organizes communications, pursues social integration, and negotiates changes. The more voluntaristic the transformation of society, the greater the centrality of the intellectual. In capitalist countries, intellectuals still remain, as a whole, ideologists and commentators. In the Soviet Union, they have been largely integrated into the political apparatus; in China, it is Mao Tse Tung who orients the party, which in turn directs the economy. This means that the intellectuals and the ideological struggle find themselves located at the very base of society.

Each one of us reacts differently to this transformation of the intellectual's position, and our feelings are generally ambiguous. One may and should speak of the destruction of *Öffentlichkeit* with Jürgen Habermas. Democracy is deeply threatened from the moment that the principal political choices are no longer the choices of society, but the choices — generally neither public nor open — of a management team chosen for its capacity to define a strategy. Technical discussions and the power relations between pressure groups replace democratic debate.

But one can interpret this evolution in a different manner: is not the progressive disappearance of the autonomy of the world of ideas the consequence of an ever more direct intervention of organized social forces? We know that a certain form of political democracy has received its autonomy from the limits of the political game, which threatens neither class domination nor the social and political exclusion of many categories. It is not the progressive disappearance of the autonomy of the political game which by itself destroys political democracy. Today as yesterday, the existence of political democracy essentially prefers pluralism, by which I mean the separation of the political, the economic, and the cultural elites; such a pluralism is even greater when a society defines itself by its performance rather than by its effort to liberate itself from backwardness and dependence.

It would be wrong to think that a technocratic society is by nature more or less democratic than another. An authoritarian regime is more totalitarian than others; a democratic regime extends the scope of liberties and oppositions more than others. But in all cases, intellectuals lose their 'autonomy' when this autonomy is linked, all at once, to the limits of political and ideological discussions, to the complexity of historical situations, and to the slowness of evolutions. Today, either intellectuals are incorporated in a social power block or are opposed to power, or they make an effort to appeal to science, or more generally to knowledge, against power and ideologies, yet knowing all the while that they cannot detach themselves completely from the social struggles.

THE DISJUNCTION OF SCIENCE
AND THE APPARATUS

The exceptional economic growth of industrialized countries during the last thirty years, associated with their extreme economic, political and ideological domination over the rest of the world, has naturally led intellectuals to accept rather largely during this period, the image of a society identified with scientific and technical progress. Did not industrial societies, in the East as in the West, show themselves capable of rapidly raising production and productivity, of consequently improving the general level of consumption, and also of creating such dreadful weapons, that it was impossible to discuss the future of the world without admitting the omnipotence of the super-powers?

The consciousness of social conflicts was thus dismembered into, on the one hand, the fascination with technical and economic progress and, on the other hand, the anxious examination of international strategies and confrontations. The scientific and technical revolution and the cold war have been the two poles — complementary rather than opposed — of the analysis of our societies. I will not return to this period, still near and yet which seems so distant, a period in which the good conscience of the elites and the ruling classes expressed themselves with a naïvety and a brutality which today seems surprising.

During this period, the essential part of social analyses dealt with the difficulties of adaptation to ever more rapid changes. They showed the slow penetration of 'modernization' in the Middle East, in Southeast Asia, and in Latin America; humanity was imagined as a caravan, in which each country had to place its feet in the tracks left by those who had preceded them on the road to progress. Psycho-sociologists were asked for means to reduce resistances to change: they replied that this could be achieved by the greater 'participation' of all. Briefly, there was a general refusal to question the direction of the course being taken and the social nature of those who directed it. The direction seemed obvious: progress, science, knowledge, etc. The managers seemed to be professionals or technicians. A perfect illustration of what I have called the sociology of gods: one does not discuss ends, only means. One does not contest God, but the church; not knowledge, but school organization; not the Prince, but the government.

Therefore, a brutal rupture is produced when, at the heart of industrial societies, appears a critique of growth itself and, much more concretely, of the identification of science, technology, and the apparatus of economic and political management.

That which is contested is the pride — the Greeks used to say *hubris* — of societies which consider themselves omnipotent and which thus indentify their capacity to transform themselves with their own performance. Societies which no longer recognize the limits of their production deify themselves and consequently sanctify their performance and, above all, their managers. It is impossible to separate these two lines of critical reflection, even if they can, at a certain moment, become disconnected and even oppose each other: on one hand, the rediscovery of the 'natural' limits of production; on the other, a refusal of the identification of knowledge and the apparatus. For the overthrow of the former image of development, and the realization that society cannot merely exploit nature, because society is also a part of nature and risks destroying itself because of its blindness, cannot be separated from the rediscovery of society's action on itself, which I call historicity, which is the stake in social conflicts and not the rationality with which managers may identify. It is this same motion which, as it rediscovers the world of nature and the world of the gods, destroys the pride of the managers and desanctifies the order which they impose by removing the guarantee of science.

This is a complex and ambiguous crisis. For it is true that this dual critique can lead us backwards. It can make us relive in a utopian manner the image of a closed society, framed into an eco-system and governed by the functional demands of its own reproduction. Such a picture is unacceptable, because a peculiar trait of human culture is the fact that it modifies its relation to its environment, that it interacts with it instead of integrating into it, as has been recalled in a very sensible manner by René Dubos in particular. At the same time, such an image can lead us to the cult of community, of equilibrium, and thus to the more oppressive forms of social and cultural integration; these weigh down on all the small communities which monastically try to live their ideal and which absorb the energy of their members in an exhausting conviviality.

But there is little chance that this crisis will encourage the utopias of the mythic return to unity, to immobility, to equilibrium. This does not mean that they do not have an extraordinary critical power, particularly when expounded by Ivan Illich. They shatter the technocratic utopia, which affirms both the omnipotence and the rationality of the ruling elites, identified with the scientific and technical revolution, whereas these elites are also, and above all, agents of the ruling power, pursuing the development of their own power, the domination of exploited categories and the repression or manipulation of oppositions — all objectives which have nothing to do with Science and Reason, and which tend more towards Machiavelli and Clausewitz than they do towards Voltaire and Einstein.

Within this large movement for the critical disjunction of

science and the apparatuses, of the forces and the social relations of production, scientists play a new role. Because they are agents of production and tend to become managers of the apparatuses of production and knowledge (and not merely notables or 'reproducers' of a traditional type of knowledge), they are placed at the heart of the debate, torn apart, and led to criticize the apparatuses and sometimes even scientific knowledge itself, a critique which is both indispensable and dangerous. Indispensable because it would be vanity to permit the belief that scientific knowledge is developed independently of the apparatuses which manage and organize it; dangerous because it risks maintaining the condemned identification of science with power.

To assert that science as such is a product of specific social and political conditions is outrageous. On the contrary, it is a question of recognizing that science invents our culture, so that it can better detach it from the grip of the apparatuses which present themselves as its exclusive agents. One never completely grasps a pure science, and one never contends with an apparatus which does not utilize science and technology, which therefore does not have internal rationality, even in the case of advertising agencies or the political police. Scholars may examine the social determinants of the development of knowledge itself, without threatening the existence of great scientific paradigms, characteristics of a culture rather than of particular social forces. After this comes an acknowledgement of the role of scientific policies and the choice of technologies. One realizes more and more that one can arrive at the same practical result by alternative paths, that is, by very different configurations of the production and decision system. We know very well that there is more than one path towards industrialization or towards the production of enough food for the world's population. By inverting our point of view, instead of presenting the evolution of mankind as a march towards progress, we should speak of the multiplicity of paths of development. This means that technological development can never be separated from political choices. A specific technological achievement, such as the production of a certain amount of energy at a certain price, or the organization of hospitals capable of treating a certain type of disease, neither imposes certain conditions on the upstream, nor does it entail certain consequences on the downstream. It is the point towards which may converge a certain number of arrangements, and from which may emanate a certain number of scenarios. One cannot suppose that there are scientific and technical forces of production which lead to a certain number of social consequences. This is an obscure idea. For we cannot understand how the study of science and technology could determine forms of social organization which would be in harmony or in discord with it. We should therefore give priority, once again to the study of socio-political processes of development and, more widely, to social change.

CONCLUSIONS

The very progress of science and technology leads to the realization that society is what it makes of itself, depending on its decisions and keeping in mind its interactions with the natural environment.

In such a society there is no longer a sacred domain which should be recognized as commanding the social order and giving it meaning. The theme of the scientific and technical revolution is of great importance, precisely because it should show this disappearance of the sacred, that is, of territories forbidden to social conflict. These may be formed in all areas of social life. But this rapid extension of the field of social action may also be translated as the sanctification of political decisions and economic interests. Those who manage society no longer serve a meta-social world of more or less distant and distracted gods; they tend to identify themselves with the meaning of history, with modernity, with rationality and, probably tomorrow, with equilibrium and integration. At the same time, adversaries are no longer rejected as sacrileges, but as irrationalists, deviants, or marginals. More than ever before power hides behind an impersonal apparatus so complex and so opaque that demands can no longer penetrate it, and the opposition withdraws into silence, hyperconformity, or on the contrary, into dream, self-destruction and aggressiveness.

To define industrialized societies by the scientific and technical revolution is at one and the same time necessary and dangerous. Necessary, because it is a specific trait of our culture to realize that social organization is produced by social action and not by the intervention of a meta-social order; dangerous, because such a formula risks hiding the fact that society's action on itself is what is at stake in the dispute between the apparatuses on one hand, and the popular forces on the other. The specific trait of the ruling classes, whoever they may be, is to identify themselves with the forces of production and the cultural models, just as churches identify with the gods and capitalism, with economic progress. From which arises the need to show, first of all, the threats to scientific knowledge, and the occasional misuse of such knowledge, for which the power bearers are responsible; and more simply, the need to recall that the state in which one finds the forces of production, thus the technological equipment, is determined by social and political mechanisms. The ensuing result is that the more science increases the capacity of societies to operate on themselves, the more societies will become different from one another. Which is exactly the contrary of the idea, until recently widely prevalent, of the progressive convergence of industrial societies. The contemporary world offers us, on the contrary, an ever increasing variety of societies, either in the process of industrialization or already industrialized. This leaves the field of sociology with almost no limits. A society is not more or less transparent to the demands of science and technology; according to a culture marked by the role of scientific knowledge, society is a system of social and political relations through which is formed a certain organization of society.

Postscript:
Two contributors to this volume discovered in my paper 'invectives' against Soviet leaders and 'some political statements not connected with science and falsifying the essence of international and inner policy of the Soviet Union.' I would like to urge the people who happen to have read the strange final remarks of Professor Richta's and Professor Fedoseyev's papers to read at least the last page of the first part of my essay, so that they can discover for themselves whether I have attacked Soviet leaders. In my own judgement the real question is: what is the latent function of these invectives against me that are certainly not connected with science?

6

CULTURAL AND SOCIAL CONTENTS OF SCIENTIFIC AND TECHNOLOGICAL REVOLUTION

Yogendra Singh
Jawaharlal Nehru University, New Delhi

The relationship between science and their social contents is indeed complex. It is involved in issues that are fundamental to the domains of culture, social structures and the social psychology of people in collectivities, ethnic groups and nations. A series of questions arise in the course of any analysis that attempts to unravel this relationship. To what extent does the scientific and technological revolution engender or harmonize with the processes of social revolution? What is the relationship between the worldview of science and the social, political and economic ideologies of human groups? Do the two ideational structures have a common ground and points of convergence? If yes, what are its social and cultural coordinates and historical conditions? If no, what implications does this dissociation have for the processes of future social development of mankind, and where does the root of this failure lie? Does it lie in the nature of the system of science and technology or in the modes of relationship between this system and the social system in general? If social systems, as historically determined structures, have the capacity to refract and differentiate the impact of scientific and technological revolutions, what then is the relationship between the inner structure of these systems — the class structure, elite composition, dominant cultural themes and styles or the symbolic systems of the society — and the processes of institutionalization of science and technology? Finally, what is the relationship between the social and economic development of nations and their systems of stratification, and the success or failure of scientific and technological revolutions?

Indeed, it would not be possible in a short essay to analyse all the ramifications of the problems that arise from the above questions. It may nevertheless be stated that the basic problem in the analysis of these problems so far has been the neglect of the issue of historicity in the modes of revolution as well as institutionalization of science and technology in different societies. This is evident if we try to define the nature of science and technology on the one hand, and on the other hand formulate the relationship between these two systems.

There is no unimodal relationship between science and technology; and whether science constitutes a universalistic explanatory principle or system is indeed now problematic in the wake of the postulates of ethno-science. Even if one leaves aside the postulates of ethno-science and ethno-methodology, and treats science as a system of universal explanatory laws, at least in some fundamental domains, the universalism of science cannot be extended to technology. Based on a single scientific principle, there are myriads of examples of the inventions of dissimilar technologies or technological systems. This occurs for two reasons: first, due to the autonomy that the invention of technological systems has enjoyed in the history of mankind over the rationalization and formulation of the scientific system and its principles. Secondly, because unlike some of the components of scientific principles, such as its calculi of abstractions, its retroductive logical system and principles of rationalization through empirical proof and disproof which are essentially universalistic in nature, the components of technological systems are capable of varied developments in several directions and forms through separate combinations of technological components or designs in the technological system.

CULTURAL CONTENTS

Moreover, the relationship between science and technology should be evaluated as a relationship between two levels of cultural systems, the universal and the historical, respectively. The historical nature of technology and its basically sociological character was noted by Karl Marx who not only identified its linkage with the species character of mankind, its evolutionary role from prehistory to history, but also established its dialectical relationship with the forces and modes of production in society. Thus, its existential and historical nature was identified. One important implication of this formulation of technology was its relativization in consonance with the levels of historical development of social structures and their processes. Interestingly, this formulation established a dialectical relationship between historicity and universalism of the system of science and technology; according to Marx, the species-character of science and technology tended to render them universalistic but their historical contextuality endowed them with specific cultural meanings and significances relative to a particular system of a society and its existential moorings. The eminently sociological nature of the Marxist conception of technology is evident from the fact that unlike the sociologists with a rational utilitarian frame of analysis characteristic of the capitalistic ideology who posit an antinomy between nature and culture, the worker and the object of his work and finally between the individual and the collectivity, Marx treats these antinomies and contradictions as mere aberrations resulting from the alienation of human labour from its true nature through institutional impediments created by the capitalistic structure of society. Marx writes:

> The practical construction of an *objective world,* the *manipulation* of inorganic nature, is the confirmation of man as a conscious species-being, i.e. a being who treats the species as his own being or himself as a species-being ... It is just in his work upon the objective world that man really proves himself as a *species-being.* This production is his active species life. By means of it nature appears as *his* work and his reality. The object of labour is, therefore, the *objectification of man's species life;* for he no longer reproduces himself merely intellectually, as in consciousness, but actively and in a real sense, and he sees his reflection in a world which he has constructed. While, therefore, alienated labor takes away the object of production from man, it also takes away his *species life,* his real objectivity as a species-being, and changes his advantage over animals into a disadvantage in so far as his organic body, nature, is taken away from him (Karl Marx 1844: 102-3 in Erich Fromm, 1963).

Technology in this sense is an extension of human labour which in an unalienated form would truly reflect the species-character of human manipulation of nature without being *against* nature. The disjunction between nature and culture thus disappears in the Marxist notion of technology through the true dialectics of labour. If the disjunction exists it is owing to the institutional malformation of society through the persistence of private property and the capitalistic mode of production. Seen in this context Daniel Bell's formulation that 'Technology in a sense, is a game *against nature,* in which man's effort to wrest the secrets from nature comes up largely against the character of physical laws and man's ingenuity in mapping those hidden paths', or that 'economic and social life is a game between persons in which forecasting has to deal with variable strategies, dispositions and expectations, as individuals seek, either cooperatively or antagonistically to increase their *individual advantage'* (italics added) (Daniel Bell, 1974; 210-11) tends to identify the problem from the narrow rationalistic value perspective. Indeed, even on this ground, Bell seems to be ambivalent since, on the one hand, he recognizes the historicist character of science and technology and on the other, he is unable to renounce his rational utilitarian worldview. He recognizes that the question of technology and science is essentially rooted into the idea of culture. He says: 'The old concept of culture is based on continuity, the modern on variety; the old value was tradition, the contemporary ideal is syncretism . . . in the radical gap between the present and the past, technology has been one of the chief forces in the diremption of social time' (Daniel Bell, 1974: 188). While his overall diagnosis that technological advancement leads to the diremption of social time is valid, his opposition between 'continuity' and 'variety' is confusing. As we have argued in another context, as traditional cultures come under the impact of modern science and technology the value system of 'continuity' of time is replaced by 'historicity' which implies the emergence of consciousness that time is not an independent variable, is not merely given, but can be manipulated, and that through technology man can master the forces of time, if not fully then indeed to a large extent (see Yogendra Singh, 1973).

Historicity is an important constituent of the cultural content of scientific and technological revolutions. In the first place it implies the autonomy of certain values, especially the fundamental or core values, in relation to the values of science and technology, which are usually of an instrumental character. Following from this it also implies that even though the postulates of science and certain systems of technology might assume a universally uniform character, their adaptations or 'mix' with the fundamental or core values of each society would differ and would lead to the development of culturally relative systems of science and technology in different societies. Sociological analyses which, in contradistinction to the above, formulate a universal evolutionary pattern for the development of science and technology derive their arguments from two types of assumptions. First, they assume that societies could be arranged into a system of hierarchy according to which the core values would essentially be derived from the dominant value orientations of the technologically advanced societies. The proponents of this theory in a sense re-echo the nineteenth-century thesis of the scholars of the 'ideology of progress' which subsequently became historically discredited. The new element in their postulate, however, is that of combining with the cultural thesis (of the superiority of the cultural models of the industrially advanced societies) the thesis of superiority of power. We find reflections of this approach in the writings of Gunnar Myrdal on the Asian situation of poverty, and at a more general level in the theoretical formulation of the thesis of 'centre and periphery' in the recent work of Edward Shils (see Myrdal, 1968; Shils, 1974).

The second assumption in the universal evolutionary model for the scientific and technological revolution is derived from what may be called the 'fallacy of over-abstraction' in determining the nature of both science and technology. In this mode of thinking the components of science and technology are identified primarily through their abstract principles, calculi of abstractions, and models; and not with concrete manifestations of the activities of science and technology in societies as objective and historically determined structures. The result is a comparison of the patterns of science and technology development in the less developed countries with those of the advanced nations, with the help of indicators which are mainly of a meta-scientific and meta-technological nature and abstracted from empirical realities. The fallacy of over-abstraction is grounded in the choice of wrong models for comparison.

It would appear that postulates which posit a universal model for scientific and technological development for all societies based on the pattern of the economically and industrially advanced nations, use arguments which are either basically culturalogical or based on abstractions which are dissociated from historical realities of the objective development of science and technology in human societies. If the former suffers from the culturalogical fallacy in its comparisons, the latter is a victim of the fallacy of over-abstraction. What indeed is required for an objective analysis of the problems of scientific and technological revolutions in the developing societies is understanding of the historicity of the cultural and structural forces and their role in the processes of these revolutions.

The limitation of the treatment of the social content of scientific and technological revolutions from purely culturalogical or abstracted frames of reference is also confirmed through the direction that such revolutions have now taken in the industrially advanced societies. These societies, which are characterized by Daniel Bell as the 'post-industrial societies', do not reflect a pattern of scientific and technological development which could be regarded as free from serious cultural and social structural contradictions. In the wake of the rational utilitarian ideology of scientific and technological revolution in these societies some major contraditions in the social content of the revolution have emerged which they are unable to receive fully. Some of these contradictions, as estimated by Daniel Bell in a futuristic sense, relate to that between the ethos of science versus politics of science, between the values of meritocracy and equality, and between plenty for the few and scarcity for the many, along with the emergence of contrived scarcity as a result of a constricted consciousness of culture. Bell writes:

> In a post-industrial society, the disjunction of culture and social structure is bound to widen. The historic justification of the bourgeois society — in the realms of religion and character — are gone. The traditional legitimacies of property and work become subordinated to bureaucratic enterprises that can justify privilege because they can turn out material goods more efficiently than other modes of production. But a technocratic society is not ennobling. Material goods provide only transient satisfaction or an invidious superiority over those with less. Yet one of the deepest human impulses is to *sanctify* their institutions and beliefs in order to find a meaningful purpose in their lives and to deny the meaninglessness of death. A post-industrial society cannot provide a transcendent ethnic — except for the few who devote themselves to the temple of science. And the antinomian attitude plunges one into a radical autism which, in the end, dirempts the chords of community and the sharing with others. The lack of a rooted moral belief system is the cultural contradiction of the society, the deepest challenge to its survival (Bell: 1974, 480).

This more than highlights the fallacy of treating the cultural model of the industrially advanced nations for the scientific and technological revolution as the appropriate and indispensable reference point for the less developed or poorer nations of the world. It also explicates the irrelevance of projecting the limited and over-abstracted culture of science and technology, comprehensible and meaningful only to a handful of the top scientists and technocrats, to the level of general societal culture and tradition. Moreover, the treatment of science and technology as a universal cultural tradition postulates a false subject-object relationship between the two systems; it is assumed that science, the rational explanatory worldview, governs technology, the mode of its rational instrumental adaptation. The logical inference is that scientific-technological revolution should reinforce the processes of social revolution in societies based on a universal evolutionary time scale, that would promote values of humanism, social justice and egalitarianism at a global level eventually.

This mode of analysis, as we have tried to show, suffers not so much from the honesty of intent as it does from the poverty of its analytic categories, especially those of culture, social structure and power. The universalistic nature of science at a certain level of abstraction cannot be denied, nor its rational humanistic role for the solution of social, material and existential problems of mankind. There is need, however, to place these ideological goals of science and technology into their proper structural and historical context.

SOCIAL STRUCTURAL CONTENTS

Scientific and technological revolution changes the coordinates of perception of reality and its structure of relationships. Not only does it revolutionize the evaluation of material and social reality through maximization of their manipulation through mass production, transport and communication, and the alteration of the perception of space and time; it brings into being new social structural forms, new professions, occupations and standards of definition and evaluation of means-end entities. This revolution introduces a new phenomenology of consciousness with dangers of its historical dissociation from the existing structure of consciousness in societies. It is said

> that for most of human history, *reality was nature,* and in poetry and imagination men sought to relate the self to the natural world. Then *reality became technic,* tools and things made by men yet given an independent existence outside himself, the reified world. Now *reality is primarily the social world* — neither nature nor things, only men — experienced through the reciprocal consciousness of self and other (Bell, 1974: 488).

The extent to which the new expected social consciousness of scientific and technological revolution gets institutionalized in societies both developed and developing would depend upon the response that the social structure of these societies makes to the challenge of this revolution.

The transitions in the phenomenology of reality from its identification through 'nature' and 'technic' to 'primarily the social world' as Bell postulates, reflects the importance of the social structural revolution for succesful institutionalization of scientific and technological revolution. Impediments in this process arise both on account of the features specific to the character of social structural revolutions and its historicity in different societies, and also due to the peculiar changes in the very nature of scientific and technological revolution at the present time. The first major issue in the nature of revolution of science and technology today is that of the inversion of their sequence and organizational form. Basically, technological revolution precedes scientific revolution in human societies. Today, however, due to near optimal advancement in the cognitive and theoretic paradigms of science and structure of scientific revolution depends more and more on technological revolutions and the potential that societies hold for organizationally and resourcefully managing these revolutions. Management of the criteria of performance and optimization of efficiency through a new type of functional rationality committed to the collective social objectives is the necessary hall-mark of the contemporary scientific and technological revolution. From this fact a fundamental situation of contradiction comes into being. The historical and sociological coordinates of promoting scientific and technological revolutions in the rich countries as compared to the poor or developing countries do not necessarily coincide; they might even run at cross purposes in terms of ideology, rationality of management of priorities and technological objectives. They might also get distorted through the nature of international relationships in the economic, political and cultural spheres, which may range from relationships of domination and neo-colonialism to that of friendship and cooperation. Under these circumstances the universalistic cultural ethos of science is subordinated to the social stratification of groups and nations and the structural asymmetries of wealth and power.

CONTEMPORANEITY AND CONTRADICTION

The structural form of the present-day scientific and technological revolution is characterized by two important features: contemporaneity and contradiction. Contemporaneity has so far been treated by most social scientists as a cultural reality rather than social structural. As a cultural theme, contemporaneity implies a new form of time-consciousness and standard of value orientations in which present orientation dominates over past orientation, belief in the possibility of transformation of life chances dominates over the value orientation of fatalism and determinism and finally, the consciousness of the reality of the 'here and now' is pre-eminent over the consciousness of a diffuse and continuous time scale. This cultural meaning of the term 'contemporaneity' apart, the concept has a deeper social-structural significance for the process of scientific and technological revolution. Contemporaneity in this context means the transformation in the scientific and technological role structures and manpower resources and their asymmetrical distribution between the nations that are industrially advanced and rich, and those that are less developed and poor. It is said that

> ninety percent of the scientists that ever lived are living today, and 90 percent of them are now in a few developed countries of the world. In addition a significant fraction of competent scientists are lost to the developed world... and an overwhelming proportion of the industrial output of the developing countries is based on technologies imported from the developed countries (NCST, 1973).

This form of operation of the principle of contemporaneity in the process of technological and scientific revolution engenders contradiction. It is generated by the very process of import of technologies. Apart from that it creates problems of balance of payments for the poorer nations on account of monopoly pricing, transfer of inappropriate technology in terms of capital-labour intensity, and it distorts the nature of priorities in production and investment in these nations. It does not help in the production of commodities at a particular level of quality which might improve the living standards of the masses and the poorer sections of society. The imported technologies in most developing nations of Asia, Africa and Latin America have succeeded only in encouraging production of 'high quality goods' to meet the requirement of the high income groups and the middle classes in these societies. This contradiction relates basically to the nature of the social structure of these developing societies in the wake of their exposure to the forces of scientific and technological revolutions. Since most of these societies have been previously colonies of one or the other developed nations of the world, and also since in most cases the colonial situation led to the creation of a middle class in these societies whose cultural role model was the ruling imperial class, the implications of political freedom for their scientific and industrial structure and its policies was highly distorted by the middle class vested interests. This has directly or indirectly governed the science and technology policies in most of these nations. Even in those nations where the middle classes were more enlightened and had revolutionary potential, the dominant economic power of the erstwhile imperial societies always tried to bend them to a new form of neo-colonialist relationship. Without a radical social revolution, as happened in China for instance, it has been difficult if not impossible for most other developing nation of the Third World to forge meaningful and effective goals for scientific and technological revolution.

This brings us to the realization of another structural contradiction of the contemporary scientific and technological revolution, especially in the developing nations whose causal linkages extend to the highly developed nations of the world. Owing to the historical dependence of the scientists and scientific institutions in the developing countries on those in the developed nations, in the light of the contradictions mentioned above, the role of these scientists and scientific institutions in generating forces of effective scientific revolution tends to be marginalized and their scientific and technological goals and aspirations are alienated from the fundamental needs of production and innovations for production in their society.

Underdevelopment not only thwarts the capacity for investment in research and development activities in the poor nations but also distorts the role of science and scientists in these societies by artificially (and not organically) linking them with the system of science and technology in the developed world, whose functional needs are of an entirely dissimilar order. Consequently, the elitism that such artificial growth of the scientific profession generates in the developing nations brings it closer to the state of neo-colonialism. This class of intellectuals and professional scientists survives on its capacity to establish patronage relationships with political, intellectual and scientific elites of the advanced nations, who are themselves often products of structural contradictions of neo-colonialism, both internal and external. It survives not on the basis of its autonomous growth in tune with the structural needs of society but through establishing its dependence on systems of technology and science dissociated from the structural capacity of their own social system to absorb the technology for social benefit and growth.

There is, therefore, a logical interrelationship between the scientific and technological revolution and social revolution in developing societies. The significance of this relationship has relevance also for the developed nations, but for the developing nations it assumes prime importance. It establishes the need for basic structural changes in the social systems of the developing countries, primarily in the orientation and organization of its elites and middle classes to engender a relevant social context for meaningful scientific and technological revolution. Evidently it would also mean that in developing nations of Asia, Africa and Latin America, the process of scientific and industrial revolution would have to proceed simultaneously with social revolution. It would imply structural changes in their system of social stratification, structure of power and elite ideologies which may promote the process of growth of science and technology for production which harmonizes with the needs of the common people and the organic interests of their societies. It would also imply the institutionalization of such systems of science and technology in these nations which do not become a disguised source of neo-colonialism and thwart the possibility of successful scientific and technological revolution in consonance with the organic and historical needs of these societies. The truly sociological and species-character of science and technology would be objectified only under these historical circumstances of social change.

REFERENCES

Bell, Daniel (1974) *The Coming of Post-Industrial Society: A Venture in Social Forecasting* (Delhi: Arnold Heinman).
Marx, Karl (1963) *The Economic and Philosophical Manuscripts* (tr. T.B. Bottomore) in: Erich Fromm, *Marx's Concept of Man* (New York: Fredrick Unger Publishing Co.).
Myrdal, Gunnar (1968) *Asian Drama: An Inquiry into the Poverty of Nations*, Vol. I and II (Harmondsworth: Penguin Books).
Shils, Edward E. (1974) *Centre and Periphery: Essays in Macro-Sociology* (Chicago: Chicago University Press).

Singh, Yogendra (1973) *Modernization of Indian Tradition* (New Delhi: Thomson Press).

Govt. of India (1973) *An Approach to the Science and Technology Plan, National Committee on Science and Technology* (New Delhi).

7

SOME POTENTIAL CONTRIBUTIONS OF LATECOMERS TO TECHNOLOGICAL AND SCIENTIFIC REVOLUTION: A Comparison between Japan and China

Kazuko Tsurumi
Sophia University, Tokyo

INTRODUCTION

Joseph Needham, in his monumental work, *Science and Civilization in China,* asserts that 'no people or group of people has had a monopoly in contributing to the development of science. Their achievements should be mutually recognized and freely celebrated with the joined hands of universal brotherhood.'[1] This attitude he calls 'a new universalism'. Science is sometimes defined as a system of knowledge, and technology as 'the organization of knowledge for the achievement of practical purposes'.[2] But science may also be defined as a dynamic process of knowing, or of making knowledge.[3] A corresponding definition of technology is the process of making things. For the purpose of this paper I prefer the second set of definitions, for two reasons. First, it avoids the ideas of sequence, causality, and consciousness implicit in the description of technology as 'organization of knowledge' for a specific purpose. Even in England, in the seventeenth century, the steam engine was developed through a series of ingenious devices worked out by technicians, mostly not educated in science. More recently, in China, acupunctural anaesthesia was found to work in some operations, but this was not the result of application of theoretical scientific knowledge: the theory is only now being investigated. Since our subject matter includes the effects of traditional technology on the processes of industrialization in China and Japan, it seems more useful to adopt definitions of science and technology as related, but not necessarily sequential, dynamic processes. Secondly, the definitions of science and technology as processes treat them as activities of men in changing circumstances, and allow us to locate where and how changes in those activities take place.

Each society has its own ways of knowing and making things, accumulated from generation to generation. These types of science and technology are expressed in the everyday languages of the peoples and are influenced by the structures of their languages. To that extent they are particular to each society and its people. There is another type of science and technology expressed in meta-language, such as in mathematical symbols, which are universally accepted and used by the peoples of different societies. Keiji Yamada calls the former type of science ethno-science to distinguish it from the latter, which is identified as modern science.[4] Modern science and its meta-languages were first developed in a few Western European countries, like England, France, and Germany; the concomitant technologies originated in Europe and in the United States at a time when there were no previous models of industrialization to fall back on.

Latecomers[5] to the process of industrialization, like Japan, China, India, and other Asian, African, and Latin American peoples, borrowed these models of modern science and technology and attempted either to impose them on, or integrate them with, their traditional ways of knowing and making things. In this process, Western science and technology comes into contact with different natural environments as well as various ethno-sciences and technologies. The resulting modifications and transformations of the original models may sometimes function as a feedback to Western models — perhaps even to create a new pattern of 'universal' science and technology, better suited to the purpose of survival of human kind.

There are two contrasting views of science and technology. The first is an optimistic view that modern science and technology are an eternal blessing, and that the evil consequences, such as pollution, can be controlled by developing more science and more technology. The other is a pessimistic one: that science and technology are the source of evils and should be stopped developing if humanity is to survive. I should like to take an intermediary position: that modern science and technology would be useful from the point of view of human survival, if and only if they are redirected by 'ethnosciences' and 'ethno-technologies' which are the pre-industrial legacies of various peoples.

I shall attempt to explore some of these possibilities. In both China and Japan, modern science and technology were first introduced in the sixteenth century, but the patterns of acceptance, rejection, or integration with traditional models have been quite different. The modern phase of development in each country showed even greater differences. I will date this phase from 1868 (the Meiji Restoration) in Japan, and from 1949 (the establishment of the People's Republic) in China.

IMITATION AND SELF-RELIANCE
AS MODELS OF DEVELOPMENT OF SCIENCE
AND TECHNOLOGY AMONG LATECOMERS

Imitation and self-reliance are not logical categories, but historically evolved concepts, which can be translated into two distinct models of development of science and technology among latecomers. In Japan, since the inception of industrialization, the imitation model was followed consistently. Since the end of the Second World War, despite the fact that export of technology is gradually on the increase, the import of foreign technology is still overwhelmingly high.[6] In contrast, China began with the imitation model during the first five-year plan period (1953-57), but notably since 1958, the self-reliance model has come to predominate. Japan and China thus appear to represent two distinct models of development of science and technology.

There are four processes of managing tensions that might arise between the newly introduced exogenous science and technology on one hand, and the pre-existing indigenous science and technology on the other.[7] The first is the monopolizing process, in which the indigenous science and technology are suppressed to the point of extinction. The second is the multi-layered process, where indigenous science and technology are allowed to persist side by side with exogenous patterns in the comparable field. The indigenous and the exogenous patterns are arranged like geological layers, compartmentalized so as to minimize confrontation with each layer used for a different purpose or sphere of activity. The third is the coexisting process, in which both indigenous and exogenous patterns are encouraged to interact, and conflict might be intensified. The fourth is the integrative process, where some efforts are made to accommodate exogenous technology and science to the indigenous situation, including both natural environment and indigenous technology and science. This process has to go through confrontation and conflict between the exogenous and indigenous elements and then to create a new pattern by restructuring them.

The imitation model is characterized by a strategy of industrialization through maximum importation of exogenous technology and science. The major consideration of this strategy is the attainment of maximum results within the minimum span of time. The self-reliance model, in contrast, is characterized by the strategy of utilizing both indigenous and exogenous sciences and technologies to industrialize the society by maximizing the efforts to design tools and contrive methods of making things best suited to the natural and social environments. This model sets relatively long-range targets to be achieved over a relatively longer span of time as contrasted with the speed emphasized by imitation model strategists. Within the framework of the self-reliance model, the same four processes of tension management are to be identified. The only differences lie in the cases first of the monopolizing process and second of the integrative process. First, the monopolizing process of the self-reliance model adopts indigenous science and technology to the virtual rejection of exogenous patterns. Secondly, since the strategy of this model sets a longer time limit for achieving its goal, this model is more conducive to a more thoroughgoing integrative process than the imitation model.

For the imitation model to be effective, the power of decision-making has to be vested in the top echelon of the central authority, whereas for the self-reliance model to be successful, decision-making power has to be diffused as widely as possible to the local level and to the members of each local community.

THE IMITATION MODEL — IRON
AND STEEL INDUSTRY IN JAPAN

The evolution of the iron and steel industry in Japan is typical of the pattern of development of large-scale industries in Japan, which went through the cycle of partially integrative and multi-layered processes, ending up with the completely monopolizing stage of relying on Western technology to the exclusion of indigenous technology.

During the Edo period (1600-1867), the indigenous method of manufacturing steel ingot and pig iron out of domestic iron sand was developed. It used charcoal mixed with iron sand in an open clay furnace into which air was fed by foot-bellows. This method was called *tatara tetsu* (foot-bellowed iron), which was mainly concentrated in the Izumo District. When the Meiji Government launched the first modern iron-manufacturing factory in the city of Kamaishi in 1875, they set aside the traditional iron-manufacturing technology, and hired a British engineer to lay out the plan and supervise the whole process strictly according to British technology. Less than one hundred days after two modern blast furnaces were lit, operations had to be stopped due to a shortage of charcoal. In 1894, the British engineer resumed the operation of the blast furnaces, by replacing charcoal by coke, which caused congelation of iron and coke in the furnaces, resulting in the closure of the entire plant. The British engineer was dismissed. Two years later, a merchant of traditional *tatara* iron goods got interested in the defunct factory in Kamaishi. With Government support he and his sons built two small new blast furnaces. Later, three small furnaces were added to make five which produced

20 percent of the iron produced in Japan, and the remaining 80 percent was still being produced in the District of Izumo in the purely traditional way. It was not until the early part of the 1920s that the indigenous iron-making workshops went completely out of operation. The period between the 1870s and the 1920s is characterized by the multi-layered process.

In 1895, under the pressure of the Sino-Japanese War, the Government asked a Japanese scholar of engineering, Kageyoshi Noro, to rehabilitate the modern blast furnaces in Kamaishi. Noro succeeded in making coke suitable for modern blast furnaces from domestic coal. It was then discovered that the failure of the British engineer, who previously made the furnaces, was due to the fact that he was not familiar with the nature of the Japanese coal. The Kamaishi plant then began to produce 65 percent of the pig iron produced in Japan. Thus it took more than ten years to rectify the situation, and it was made clear that the immediate application of foreign technology is bound to fail, unless at least some accommodations are made to the local circumstances. In 1902 the second and largest iron and steel plant was constructed by the Government in Yawata. This time technology was borrowed entirely from Germany and a German engineer was employed to plan and supervise it. Iron ore was imported from China, but domestic coal was still used to make coke. After a year's operation, the plant developed defects similar to those of the Kamaishi factory, and again Noro was called in to resolve the problem. These two episodes exemplify the partially integrative process.

Finally in 1906 the Government decided to import coal from China to step up production, which completed the cycle of making Japan's iron and steel manufacturing industry entirely dependent upon foreign technology and raw materials.[8] The Yawata Iron and Steel Plant in 1970 ranks third in the world in annual production of splinter steel, relying solely upon foreign technology for the entire manufacturing procedure.[9] This last stage represents the monopolizing process.

Large-scale industries in Japan are more daring about grabbing the most advanced technology from abroad than about investing their funds in basic research to cultivate their own technology. In 1972, Japan's production of splinter steel ranked third in the world. Japan boasts the largest and the second largest steel plants in the world, and the highest percentage of use of LD type revolving furnaces. The LD converter is the most advanced technique for converting pig iron to steel ingot, much more efficient than the conventional open-hearth furnace.[10] But due to the intensity of the heat at which the pig iron is melted, it also emits the most oxidized iron dust into the air. Seventy percent of the steel plants in Japan, including the world's largest and the second and fourth largest ones, are concentrated in the area surrounding the Seto Island Sea.[11] The bigger and the more concentrated the plants, the more dust, smoke, and sulphurous acid and other gases are likely to be emitted in to the air. The Japanese industry accentuates the organizational principles of the Western industrial society, namely, the gigantization of units and the concentration of big industrial complexes in urban areas, thus aggravating the effects of pollution. Consdiering the relative smallness of the geographical size of the country, the adverse effects are enormous. However, paradoxically enough, the situation is responsible for creating novelties in the scope of anti-pollution activities among local inhabitants, from which germinate some new views of science, technology and social organization that sustain them. This point will be discussed in the last section of the paper.

FROM IMITATION TO SELF-RELIANCE –
IRON AND STEEL AND OTHER
INDUSTRIES IN CHINA

China also began by attempting to impose borrowed technologies. A switch from the imitation strategy to that of self-reliance in 1958 was marked by a campaign mainly in the agrarian areas for 'backyard furnaces' based on indigenous technolgy *(tu-fa)*.[1,2] The year 1958 also marked an upsurge, among the agrarian masses, of movement toward reorganizing their cooperatives into people's communes. Under these circumstances the slogan 'self-reliance' *(Zu-li geng-sheng)*[13] had a double meaning: to make people's communes independent from reliance upon abroad, and at the same time to make them self-sufficient in the context of national economy. In order to do so, each commune should be able to produce not only food but also the daily necessities of its members. To achieve such an end backyard furnaces were chosen, because they were relatively small and simple so that peasants could easily learn to construct and operate them to produce iron and steel with which to make their own farm implements and household utensils. The backyard furnaces failed to produce iron and steel of decent quality, but although the experiment was short-lived, it was significant in producing new attitudes toward technology and social organizations to develop it.

The Chinese 'self-reliance' model[14] is based upon a long-range program of **urbanizing** villages and at the same time agrarianizing cities so that either unit may become self-sufficient, and polarization between cities and villages may be avoided.[15] In contrast to foreign machines which have been characterized as 'gigantomaniac', Chinese machines are smaller in size, lighter in weight and simpler in structure.[15] These characteristic features made the Chinese machines fit for use by an agrarian population with rudimentary knowledge and training. The light-weight, simple-to-operate machines are also suitable for dispersing small- and medium-sized industries to communes to make them self-sufficient. These two trends are clearly deviating from the Western model of industrialization, in which urbanization is a 'universal' concomitant, and technological innovations are expected from, and only from, experts in the university, government, and big-business laboratories.

During the periods of the Great Leap Forward (1958-61) and the Great Cultural Revolution (1966-71), the policies of 'walking on two legs' and 'three in one' were implemented in full. The first policy advocated the simultaneous development of indigenous and modern technologies, central and local industries, small-, medium-, and large-sized industries, and heavy and light industries with agriculture as their base.[16] The second policy prescribes the populist basis of innovators. It advocates that workers and peasants, engineers, and cadres should work closely together so that the indigenous technologies of the workers and peasants are brought into confrontation and combination with modern technology and expertise.

An abundance of incidents of technological innovation are cited both by native and Western observers.[17] They range widely from very simple and down-to-the-earth improvements like the use of marsh gas generated from human and cattle manure to be piped to individual household kitchens for fuel and light,[18] to highly technical innovations, such as the development of synthesized insulin in 1965.[19]

In the area of agro-technology, for instance, a Canadian diplomat finds, in a commune in the Pearl River Delta, a wooden machine to lift 'water from the river for the cultivation of rice'. This 'simple but ingenious' device was constructed by cooperation of peasants and a few engineers from a university. This machine also generated power for small factories and electricity for household use.[20]

THE SELF-RELIANCE MODEL — TOWARD INTEGRATION OF CHINESE AND WESTERN MEDICINES

In China, medicine is the most impressive field in which indigenous and Western sciences and technologies are coming into a novel and successful union.

In Japan, on the other hand, the usual multi-layered type of arrangement occurred. Chinese medicine was the accepted practice in Japan until knowledge of the 'Dutch learning' began to infiltrate toward the end of the Tokugawa Exclusion period. In 1875, shortly after the Meiji Restoration, the Ministry of Education declared that all doctors must pass an examination in Western medicine. Those already in practice were exempted from the rule, but there were to be no future physicians specializing in traditional Chinese medicine. The traditional doctors' groups petitioned the Parliament without avail.[21] But acupuncture, moxibustion and herbal medicine survived as folk therapies. Although these therapies are still popular in Japan today they are jealously separated from medicine proper which is monopolized by Western technologies.

In China, the history has been very different. Western medicine was introduced into China by missionaries in 1827. Traditional medicine and Western medicine came into severe conflict. At the time when the Kuomintang Government in 1929 attempted to eradicate the traditional doctors by announcing 'the abolition of old medicine to clear away obstacles to medicine and public health', the number of traditional doctors exceeded that of Western medical doctors and the former were strongly supported by the people, especially in the villages. The organized opposition of Chinese doctors and their supporters was so potent that the Government had to repeal its resolution. Subsequently, the Institute for National Medicine was established by traditional doctors and by scholars who supported them. This institute advocated 'scientification of Chinese medicine'. From that time traditional versus modern medicine became a controversial issue not only among doctors but also among intellectuals and politicians. In 1943 the Kuomintang Government issued the 'Physicians' Law', according to which the traditional doctors were entitled to acquire their doctor's licence either by obtaining a diploma from Chinese medical schools, by having practised for five years with good reputation, or by passing the Government Examination. Thus they were given the same privileges as modern physicians,[22] and they maintained roots among the people, owing to an acutely short supply of modern physicians. This was especially true during the war of resistance against Japan in agrarian base areas of the red army. In the base areas, medical treatment was in short supply, and whatever was available, traditional and modern, had to be put to the best possible use.[23] Experiences in these situations of dire need provided prototypes for the later development of collaboration between both types of medicine. At the time of the establishment of the People's Republic of China in 1949, there were only 10,000 licensed modern physicians[24] for 574.2 million population,[25] but there were 500,000 traditional doctors.[26]

In 1950, four basic principles of national health were adopted by the National Health Congress: (1) 'to look toward the workers, peasants and soldiers'; (2) to emphasize preventive medicine; (3) to unite Chinese and Western medical doctors; and (4) to combine health care with mass movements. These programs were implemented during the Great Leap Forward and the Great Cultural Revolution. Three prominent features emerged in the development of medicine. First, the system of medical training was redesigned to make the mastery of Chinese medicine a requirement for modern medical doctors. Secondly, a combined program of Chinese medicine and Western medicine proved effective in the massive campaign to eradicate endemic diseases among the agrarian population. In this campaign, agrarian people participated in the practice of preventive medicine through their efforts to make their environment inhabitable. Thirdly, acupunctural anaesthesia proved to be effective in various operations. This stimulated doctors in China, and eventually abroad, to probe for a theory to account for acupunture.

At the end of 1954, a new program was intitiated under which 120 high-caliber modern doctors were chosen to study Chinese medicine under thirty veteran Chinese medical doctors for two years, plus one year of clinical experience. Through this new program, it was expected, modern physicians would not only master traditional medicine, but also endeavour to test its effectiveness and systematize its theories through their scientific method.[27] By April 1960, the number of modern physicians who became full-time students of Chinese medicine increased to more than 2,300. At the same time, Chinese Medical Schools of high and middle levels, and Schools for Improvement of Chinese Doctors were reinforced. In those schools, although emphasis was put on Chinese medicine as the foundation, the study of the minimum essentials of modern medical knowledge of anatomy, physiology, pharmacology and bacteriology was required. Outside of those regular schools, the traditional system of apprenticeship was encouraged to transmit Chinese medical skills to the younger generation. In June 1958, there were approximately 68,000 registered apprentices.[28] By 1956, it was established as a rule for all the modern medical schools to include Chinese medicine as an integral part of their curriculum.[29]

Acupunctural anaesthesia was practised for the first time in medical history in 1958 in a Shanghai hospital. During the Cultural Revolution it came to be in prevalent use, and since then 400,000 operations, including tonsillectomies, pneumonectomies, thryroidectomies, and brain surgery among others, have been performed. Nearly 90 percent of them are reported to have been successful.[30] Dr K.K. Jain, a Canadian physician with the Western medical study group in China, claims these advantages for acupunctural anaesthesia: (1) the patient is kept awake and is able to cooperate with the surgeon during the operation; (2) there are no after-effects such as nausea and vomiting; (3) recovery is quicker than in the case of operations with general anaesthesia, and consequently there are less post-operative complications; and (4) no death has yet occurred with acupunctural anaesthesia, whereas with general anaesthesia death rates are two per 100,000 operations.[31] In spite of these advantages, acupunctural anaesthesia has as yet been very little used in the West.[32] The slowness of recognition of acupunctural anaesthesia by Western medicine is due to the lack of any conclusive theoretical explanation of how and why needles pressed into certain points of a human body produce analgesia at other parts of the body.

ETHNO-SCIENCE AS THE BASIS
FOR COPING WITH POLLUTION

Pollution is defined by some biologists as 'the general deterioration of the environment upon which humanity is completely dependent for life'. It is caused by man-produced substances with which '*Homo sapiens* has no evolutionary experience and against which human cells have evolved no natural defense'. These substances may affect man directly as in the case of inhaling polluted air, drinking polluted water, and eating polluted food. They may also bring adverse effects on him indirectly by undermining the *ecosystems*, 'the functional relationship among organisms and between organisms and their physical environments', as in the case of deforestation, radioactive pollution, and the use of synthetic pesticides and herbicides.[33]

In 1956, the Chinese government proposed the principle of 'multi-purpose' use of industrial wastes. According to this principle, waste liquid, waste heat, and waste solids are recycled to produce new raw materials for further production and at the same time to prevent the former from becoming pollutants. In Shanghai, for instance, where industrial plants are most highly concentrated in China, coordination in utilizing industrial wastes is practised. A resin factory previously used sulphuric acid and hydrochloric acid to produce chloromethyl methyl ether. But by coordinating their production with a chemical factory, the former is now supplied with a certain type of waste gas from the latter to produce the ether, thus saving a great quantity of sulphuric and hydrochloric acids.[34]

In agriculture, human and cattle manure not only in the villages but also in the cities is brought back to the farms, and after being treated it is intensively utilized.[35] There is also a new trend for research and production on the local level of bacterial fertilizers and microbe insecticides.[36] Combined with human and animal manure and microbe fertilizer and insecticides, the use of chemical fertilizer may be kept at a relatively low level, thus preventing soil and food poisoning. The Chinese ways of recycling both industrial and biological wastes are very close to what modern ecologists are recommending to prevent pollution and at the same time conserve the soil and natural resources.

However, the Chinese are more successful in implementation than most of the industrially more advanced societies. Three reasons may be pointed out. First, the higher the general level of industrialization the more difficult it is to cure and prevent pollution. China is at the stage of industrialization where long-range preventive measures have better chances of success. Secondly, the recycling of industrial wastes involves high costs and does not yield quick profits, and thus it conflicts with the principle of maximization of profits in private enterprise systems. The Chinese collectivist orientation and calculation of balance sheets in long-range perspectives are more conducive to a process that does not yield quick profits but assures a better environment to the community as a whole.

Thirdly, recycling is compatible with Chinese ethnoscience and technology. J.B.R. Whitney, a Canadian geographer, characterized Chinese traditional agriculture as 'partial imitation of natural ecological processes'.[37] He explains the traditional agrarian urban symbiosis as follows:

> The stability of these local farming ecosystems was maintained by a highly organized system of spatial nutrient cycling. Nutrients in crops exported to the local market town in the form of rent, taxes, or produce for sale were, in part, returned to the locality of origin by means of a flourishing nightsoil trade.[38]

This traditional agrarian recycling model again was based

upon the cyclical view of the universe of the ancient Chinese, derived primarily from the observation of the life processes of plants in accordance with the changes of seasons.[39] Japan's problem of pollution differs from the Chinese case at least in two ways. First, Japan is far more advanced than China in the degree of pollution and in the variety of pollutants. Secondly, in China pullution-preventive measures have been encouraged by the political leadership, whereas in Japan it is the local citizens and their voluntary associations that constitute the major forces to protect themselves from polution. They are confronted with business and administrative elites who are responsible for neglecting and aggravating environmental disruption. The elites are committed to the imitation model, whereas the people are evolving forms of the self-reliance model in their struggle to cope with pollution.

Historically, the present-day Japanese have two groups of prominent forerunners in anti-pollution movements: the embittered and hopeless fight against the pollution of the Watarase River and its basin by the Ashio Copper Mine owned by Furukawa Mining Industry; and the fight against the destruction of the ecosystems resulting from local government officials' zealous efforts to implement the central government's order to merge native Shintoist shrines. Both incidents occurred in the early stage of industrialization in Japan, but they set the patterns of destruction by industry and by the government, and established the models of confrontation on the part of the local inhabitants. The Ashio Copper Mines was started in 1878, and by 1885 it produced 39.2 percent of all the copper in Japan. The copper slag and mineral gases resulting from mining and smelting poisoned the river water. This went on unrecognized until the flood of 1891 hit the rice fields along the river, making them barren, and also killing the fish in the river. By 1892 there were no more fishermen along the river. Another flood in 1897 victimized 119,000 households and about 520,000 inhabitants. Tanaka Shōzō, who was elected to the parliament in 1891, campaigned persistently and vigorously in the parliament for ten years to persuade the government to suspend the operation of the mine. In 1898, the victimized peasants organized the first petition to the government to stop Furukawa polluting the area and jeopardizing their lives. Instead of ordering the mine to cease operations, the government advised Furukawa to take some preventive measures. The peasants' protests persisted until finally they were suppressed and many of them were arrested in 1900. In June 1907, under the government order issued by Internal Affairs Minister Hara the entire village of Yanaka, where the damage was severest, and the inhabitants were most vocal in protest, was demolished. However, sixteen households were determined to remain, and Tanaka Shōzō, having left the parliament as an act of protest, entered the village and lived among the villagers, sharing their hardships for twenty years.[40]

Tanaka Shōzō was the leader of the first anti-pollution movement in modern Japan, and he was the first person who used the word *kōgai* (public nuisance or pollution).[41] In his accusation of the government which demolished Yanaka village he stated very clearly that the members of the village had the inalienable right of self-determination, which was based firmly upon the rights to own and cultivate the land handed down from their ancestors. He asserted that the illiterate, destitute, homeless peasants who protected their ancestral land and defended their right to autonomy were all *kami* (gods).[42]

The second example of an anti-pollution movement dates back to 1906. Since 1888, merging of villages and towns had been in progress to reduce their number to one-thirtieth of what it was during the Edo period, and thus make it easier for the central government to control them. Previously, each village had at least one folk Shintoist shrine. After merging took place more than two shrines came to exist in one village or town. Thus the government considered it only logical that shrines also should be merged. The destruction of shrines, and their surrounding forests, was especially devastating in the Prefectures of Mie and Wakayama, where Minakata Kumagusu, in internationally known microbiologist and folkorist, happened to live. By 1911, in Wakayama, the number of the shrines was reduced to one-sixth of what had previously existed. Minakata, concerned about the ecological and social effects of the devastation, wrote petitions in 1911 to influential men in Tokyo. In his petitions, he wrote how the swallows and many other birds which used to make nests on the eaves of old shrines and in the forests had disappeared since the destruction of the shrines and deforestation, and this caused a radical increase of insects that damaged crops on the farms. Consequently, farmers were forced to buy insecticides, which made for increased cost of production. Since the forests on the waterfront were also cut down, the fish that used to come to rest under the shadows of trees stopped coming to the shore. This made it necessary for fishermen to switch from inshore to offshore fishing, for which they were not equipped. As a microbiologist, he also deplored that with plants and foliage being destroyed, rare specimens of microbes might become completely extinct. He continued his campaigning with local farmers, fishermen and craftsmen for ten years, until finally the decree was repealed in the House of Peers in 1920.[43]

Minakata was a man of science, and developed his antipollution argument on an ecologist basis. At the same time we detect a parallelism between his ecological reasoning based on the symbiosis of living creatures with their environment and his belief in the co-existence of various kinds of religious truth. Both Tanaka and Minakata had these views in common: first, that in accordance with indigenous folk Shintoism, all living beings are endowed with spirits whose functions may differ, but all are of equal value; secondly, that the members of a local community should have the right to decide its fate.

The Japanese are suffering from a variety of major sources of pollution: Yokkaichi asthma caused by the air polluted by the concentration of large petrochemical industries; ouch-ouch disease caused by cadmium discharged from mining and smelting plants into the Jinzu River in Toyama Prefecture; Minamata disease caused by the dumping of methyl mercury into Minamata Bay by the Chisso chemical factory; Kanemi Rice Oil poisoning caused by the intake of PCB-contaminated oil; hazardously high accumulation of PCB in a variety of fish, which is the major source of protein for the Japanese; equally hazardous accumulation of BHC through its use in insecticides and in drinking water, milk, vegetables and beef. To cope with these adversities, more than a thousand local citizens' voluntary associations have been formed all over the country.[44]

In these local citizens' movements, intimate cooperation among victims of pollution, the wider public, and scientists is emerging. Those scientists involved in the movement feel it necessary to listen to the experiences of the victimized persons and to identify themselves as much as possible with them, in order to probe into the intricate causes of disease, which are not yet fully known to present-day science, and whose elimination is not yet adequately studied by present-day technology. From among these scientists a new approach to science and technology is burgeoning. Jun Ui, biochemist and a leader of the anti-pollution movement, emphatically asserts: 'It is indeed very foolish to dissipate the gifts of nature for the sake of economic over-development. A theory that people and nature come first has not yet been systematized, but there is some hope that such a theory will be established in the furture.'[4][5]

An attempt to rectify the adverse effects of industrialization does not lead us to the denial of modern science and technology, but to a recognition of the relevance of preindustrial legacies of ethno-science and technology. Since latecomers are less far removed from the pre-industrial period, they are in a better position to have some of those legacies kept intact. The Chinese have their cyclical theory of man and nature by which to redirect modern science and technology. In spite of their achievement in very rapid industrialization, the Japanese have retained the primitive belief in the symbiosis of nature and man, with which they can reorient the modern science and technology they have acquired from the West. If there is to be any scientific and technological revolution that is meaningful from the point of view of human survival, it must come of such recreation of modern science and technology through integration with the ethno-science and technology which peoples have inherited from their pre-industrial past. And that new science and technology must be made compatible with decentralized structures of power.

NOTES

Deep appreciation is due to Ms. Judith Merril and Professor Jay Macpherson of the University of Toronto for their generous and invaluable assistance in editing my manuscript. All Japanese names in the text appear with the surname first and the given name second.

1. Joseph Needham, *Science and Civilization in China,* Vol. I (Cambridge University Press, 1961), 9.

2. Emmanuel G. Mesthene, *Technological Change: Its Impact on Man and Society* (Harvard University Press, 1970), 25.

3. Charles Singer, *A Short History of Scientific Ideas to 1900* (The Clarendon Press, The University of Oxford, 1962).

4. Yamada Keiji, 'Pataan·Ninshiki·Seisaku-Chūgoku no Shisōteki Fūdo' (Pattern, Recognition, and Making — Chinese Intellectual Climate), in Tōru Hiroshige, ed., *Kagakushi no Susume (An Invitation to History of Science)* (Chikuma Shobō, 1970), 85-87.

5. The concepts of 'firstcomers' and 'latecomers' are borrowed from Marion J. Levy Jr, *Modernization: Latecomers and Survivors* (Basic Books, 1972), 4, n. 2.

6. A Table of Import and Export Technology in Terms of Money Invested (in million dollars).

	Japan			USA		
Year	Export (A)	Import (B)	A/B	Export (A)	Import (B)	A/B
1963	9	136	0.07	927	111	8.35
1964	14	156	0.03	1,057	127	8.34
1965	17	167	0.10	1,246	134	9.29
1966	19	172	0.10	1,380	140	9.86
1967	27	239	0.11	1,567	171	9.16
1968	34	314	0.11	1,805	194	9.30
1969	46	368	0.13	1,858	192	9.68
1970	59	433	0.14	2,158	227	9.51
1971	60	488	0.12	2,465	218	11.31
1972	74	572	0.13			

England				France		
Year	Export (A)	Import (B)	A/B	Export (A)	Import (B)	A/B
1963				138.6	188.7	0.73
1964	123.2	116.1	1.07	144.0	191.0	0.76
1965	133.8	128.5	1.04	168.0	213.0	0.79
1966	160.1	132.4	1.21	180.0	243.0	0.74
1967	175.5	164.6	1.07	195.0	230.0	0.35
1968	204.5	185.0	1.11	164.5	275.2	0.60
1969	211.9	212.4	1.00	193.3	305.5	0.63
1970	263.5	239.3	1.10	214.4	349.9	0.61
1971	282.7	264.7	1.06	263.8	450.3	0.59

West Germany			
Year	Export (A)	Import (B)	A/B
1963	50.0	135.3	0.37
1964	62.0	153.3	0.40
1965	75.3	165.5	0.45
1966	73.3	175.3	0.42
1967	89.8	192.0	0.41
1968	98.5	219.5	0.45
1969	96.5	251.3	0.38
1970	118.6	307.1	0.39
1971	148.9	405.2	0.37
1972	209.2	465.5	0.45

The Office of Science and Technology, *Scientific and Technological White Paper* (The Government of Japan, 1974), 284.

7. The concept of 'society as a tension-management system' is derived from Wilbert E. Moore, *Social Change* (Prentice-Hall, 1963), 10-11. The typology of tension management is originally formulated in Kazuko Tsurumi, *Social Change and the Individuals: Japan Before and After Defeat in World War II* (Princeton University Press, 1970). See also Kazuko Tsurumi, *Kōkishin to Nihonjin: Tajūkōzōshakai no Riron (Curiosity and the Japanese: A Theory of the Multi-layered Society)* (Kōdansha, 1972). The four types of tension management apply to any kind of tensions, arising from unequal allocation of power, income, prestige, ideal versus actual discrepancies, conflict of ideologies, etc.; however, discussion is limited here specifically to the

tensions between exogenous and endogenous science and technologies.

8. For the evolution of the iron and steel plant in its initial stage, see Okumura Shoji, *Koban, Kiito, Watetsu (Coin, Silk, and Endogenous Iron)* (Iwanami Shoten, 1973), 127-209.

9. Hoshino Yoshirō, *Hankōgai no Ronri (The Logic of Anti-pollution)* (Keisō Shobō, 1972). 38; Hoshino, *Nihon no Gijutsukakushin (Technological Innovation in Japan)* (Keisō Shobō, 1966), 43.

10. The Comparative Percentages of Splinter Steel Produced by Different Methods of Production by Countries.

| | Open hearth furnace | Revolving furnace | | Electric furnace |
		LD type	Others	
		1960		
Japan	68.8	11.9		20.1
USA	87.0	3.4	1.2	8.4
England	84.5	0.5	8.1	6.9
West Germany	47.2	2.5	43.9	9.4
		1970		
Japan	4.1	79 1	-	16.8
USA	36.6	48.2	-	15.2
England	47.2	28.1	4.2	19.5
West Germany	26.2	55.8	8.2	9.8

A survey made by the Japanese Association of Iron and Steel Enterprises, quoted from Hoshino, *Hankōgai no Ronri*, op. cit., 18.

11. Hoshino, ibid., 38.

12. There is an evolution of the meaning of indigenous technology *(tu-fa)* in the Chinese context. First, it simply meant the traditional way of making things, as exemplified by 'backyard furnaces', which mushroomed all over the countryside at the beginning of 1958. Secondly, it came to mean any method of production practised by the people of a certain locality, catering to the peculiarities of that locality, making use of the local products for its raw materials, and operating on a relatively simple device. Thus in this second stage it means any locally based technology. Thirdly, the content of indigenous technology evolved in such a way as to include all the cumulative efforts

of the people to improve their own ways of making things, utilizing the best available resources, both endogenous and exogenous. In this last sense, indigenous technology became the symbol of the self-reliance model. Yamada Keiji, 'Igaku ni okeru Dentō karano Sōzō' (Creation out of Tradition in Medicine), *Tenbō*, May (1974), 48.

13. *'Zu-li geng-sheng'* is the expression used by Mao Tse Tung in his writings. In one place, it is translated as 'self-reliance' and in another place, as 'regeneration through one's efforts'. 'We Must Learn to Do Economic Work' tells: 'We Stand for *self-reliance*. We hope for foreign aid but cannot be dependent on it; we depend on our own efforts, on the creative power of the whole army and the entire people' (10 January 1945) *Selected Works*, Vol. III, 241 (author's italics). 'The Situation and Our Policy after the Victory in the War of Resistance against Japan' proclaims: 'On what basis should we rest on our own strength, and that means *regeneration through one's own efforts* (13 August 1945) *Selected Works*, Vol. IV, 20 (author's italics).

14. These long-range programs were laid down in 1956 in two programs: 'A Long-Range Plan for the Development of Science and Technology: and 'A Program for Development of National Agriculture'. These programs came to be implemented after 1958. The detailed account of the content of the programs was reported by Yamada Keiji on 2 October 1971 to the Study Group on Reconsideration of the Theory of Modernization, the Institute of International Relations, Sophia University.

15. E.L. Wheelwright and Bruce McFarlane, *The Chinese Road to Socialism* (Monthly Review Press, 1970), 167.

16. The program was stated at the Eighth National Conference of the Chinese Communist Party in May, 1958, quoted from Okazaki, *Chugoku no Tekkōgyō to Kikaikōgyō no Gijutsu Suijun (The Technological Levels of Chinese Steel and Machine Industries* (Ajia Keizai Kenkyujo, 1962), 235.

17. The Chief of the National Planning Committee, Bao Yi-bo reported that between January and 10 March 1960, incomplete statistics of twenty-four *xian* reveal that there were 25,300,000 items of technological innovations proposed, out of which 9,650,000 were adopted. These consisted of innovations in tools and facilities, production devices, waste saving and waste reclaiming. Okazaki, ed.,

op. cit., 252.

18. Mary Sheridan's paper delivered at the Annual Meeting of the Canadian Society for Asian Studies, held on 1 July 1974. The title of her paper was 'Change and Development in Rural North China and Szechuan', and the data are based upon her field research conducted in Chinese communes in the summer of 1973.

19. Joshua S. Horn, *Away with All Pests; An English Surgeon in People's China: 1954-1969* (Monthly Review Press, 1969), 111-12.

20. Chester Ronning, 'I Never Thought I Would Hear A Chinese Scholar Boast about Shovelling Dirt', *The Globe and Mail Weekend Magazine*, 20 July (1974), 18-19.

21. Akira Ishiwara, *Nihon no Igaku (Medicine in Japan)* (Shinbundō 1959), 187-89, see also Takeyoshi Takahara, *Dochaku no Gakumon no Hassō (The Pattern of Thinking of Indigenous Learning)* (Tōyōkeizai Shinbunsha, 1973), 21-22.

22. Ralph C. Croizier, *Traditional Medicine in Modern China: Science, Nationalism, and the Tensions of Cultural Change* (Harvard University Press, 1968), 90-96.

23. 'Our hospitals up in the mountains give both Chinese and Western treatment, but are short of doctors and medicines. At present they have over 800 patients. The Human Provincial Committee promised to obtain drugs for us, but so far we have received none.' 'Struggle in the Jinggnag Mountains', *Selected Works of Mao Tse-Tung*, Vol. I (Foreign Languages Press, 1967), 83.

24. Yamada, 'Creation out of Tradition in Medicine', 35; Croizier, op. cit., 157-58.

25. According to the census taken in 1953. Y. Akino and S. Kobayashi, *Hirakeyuku Chūgoku Keizai (Developing Chinese Economy)* (Tsūshō Sangyō Chasakai, 1971), 19.

26. Both Yamada and Croizier agree on this number of the traditional doctors.

27. Yamada, 'Creation out of Tradition in Medicine' 34-35.

28. Croizier, op. cit., pp.179-80; Yamada, ibid., 39-40.

29. Yamada, ibid., 45.

30. K.K. Jain, *Health Care in New China* (Rodale Press, 1973), 99-103.

31. Ibid., 109-12.

32. Ibid., 112-14.

33. Paul R. Ehrlich and Anne H. Ehrlich, *Population, Resources and Environment* (W.H. Freeman and Co., 1972), 145-46, 193.

34. Ji Wei, 'Turning the Harmful into the Beneficial', *Peking Review*, 28 January (1972), 5-7.

35. That the traditional type of fertilizer is intensively used is confirmed by a private conversation with Professor Mary Sheridan.

36. 'Widespread Use of Micro-Organisms', *Peking Review*, 4 February (1972), 11-12.

37. J.B.R. Whitney, 'Ecology and Environment Control', in: Michel Oksenberg, ed., *China's Developmental Experience* (Praeger Publishers, 1973), 96.

38. Ibid., p. 98.

39. Yamada Keiji, *Shushi no Uchūkan (Zhu-zi's View of the Universe)*, Vol. 37 (Tōhogakuhō, 1966) 131-33.

40. Kanson Arahata, *Yanakamura Metsubōshi (The History of the Annihilation of Yanaka Village)* (Shinsensha, 1970); Takeiji Hayashi, 'Watarase Gawa Kōdoku Jiken to Tanaka Shozo' (The Mine Pollution Incident of the Watarase River and Shozō Tankaka), *Shisō no Kagaku*, April-May (1971); Toyonori Okabe, 'Mine Pollution', Ui Jun, ed., *Polluted Japan* (Jishū-Kōza, 1972), 48-49.

41. Hayashi Takeji, 'Bōkoku Nihon Karano Saisei o Motomete' (In Search of Rebirth out of Japan, the Ruined Country), *Tenbō*, December (1972), 85.

42. Hayashi Takeji, 'Yanaka Mura no Fukkatsu' (The Revival of the Yanaka Village), Sōzō Shimada, *Tanaka Shōzō Ō Yoroku (The Gleanings of Old Mr Shōzō Tanaka)* (San-ichi Shobō, 1972), 513-548.

43. *Minakata Kumagusu Zenshū (Collected Works of Kumagusu Minakata)*, Vol. 7 (Heibonsha, 1971), 477-594; see also, Kazuko Tsurumi, 'Chikyu Shikō no Hikakugaku' (A Globe-Oriented Comparativist), ibid., Vol. 4, 605-14.

44. Ui Jun, ed., op. cit., 13.

45. Ui Jun, 'Pollution in Japan — A General Over View', Ui Jun, ed., op. cit., 13.

8

ON THE RELEVANCE OF HISTORICAL MATERIALISM IN THE EMPIRICAL STUDY OF ADVANCED CAPITALIST SOCIETIES

Ulf Himmelstrand
Uppsala University, Sweden

No social order ever disappears before all the productive forces for which there is room in it have been developed; and new higher relations of production never appear before the material conditions of their existence have matured in the womb of the old society. (Karl Marx, *A Contribution to the Critique of Political Economy*).

When are material conditions mature enough for the appearance of new higher relations of production? Amazingly little systematic and scientific analysis has been directed to this question and to the precise meaning of the notion of 'maturation of material conditions for the existence of higher relations of production'. In retrospect it is easy to say that the time was ripe, and the conditions mature, for societal transformations, once such changes have taken place – even though some sceptics may have doubts about the timing as well as the content even of past changes. Prospectively, more often than not, the answer to the question how mature conditions are for drastic societal changes would seem to be determined on the basis of intuitive political expediency alone. However, such 'answers' beg the question.

It would seem that at least the following conditions must be fulfilled before we can properly say that the time is becoming ripe, and that material conditions for new social relations of production have matured within the womb of the old society. Given the existence of obvious contradictions between forces and relations of production, and the 'fact' that all the productive forces for which there is room in the existing social order have been developed, three further conditions might be considered prerequisites for the appearance of 'new higher relations of production':

(1) the strength of demands for such societal change;
(2) the relative weakness of resistance to such change; and
(3) capabilities of implementing socialism at levels of popular forces, political decision-making and government (state) administration.

This very general proposition contains a number of terms which need further conceptual clarification. To the best of my knowledge Marx did not clarify his very general statement that 'all productive forces for which there is room in it have been developed' in his later work. We are thus free to make our own interpretations.

One way of interpreting Marx on this point implies a quantitative approach. It might be said that the development of productive forces possible within, say, a capitalist industrial order, has reached its limits when industrial mass-production technology and industrial work, and the market organization of commodity and service distribution *completely permeate* the given society, with nothing but insignificant segments or enclaves of other modes of production remaining or emerging. Further elaboration of this 'quantitative' type of interpretation would take account of the *type of technology and manpower* involved — capital intensive automation with accompanying unemployment bringing us even closer to the 'limits' of the development of productive forces within the capitalist order.

A somewhat different, perhaps more dialectical approach, would be to interpret these limits in terms of the adjustments made by the dominant class to accommodate and absorb the strains and contradictions generated by the development of productive forces within that order. For instance, if a capitalist class and capitalist management find that they lose more money from labour unrest and wildcat strikes than they gain from their exploitation of industrial labour, then a 'limit' has been reached, and social-welfare measures as well as more demanding trade unions may be accepted in order to reach a new, 'higher' level of accommodation. If even these measures turn out to be insufficient, as indicated by the growing difficulties to recruit sufficiently trained labour to the depersonalizing and alienating assembly lines, then another 'limit' has been reached, and industrial leadership is forced to introduce other measures like a new technology suited to the kind of 'self-managing' work groups now introduced by Saab and Volvo. If even that accommodation is insufficient and workers now clamour for more immediate control of the means and management of production -- that could possibly be the ultimate limit suggested by Marx. It would seem obvious, however, that different limits turn out to be 'ultimate' under different historical and cultural conditions. Workers in some societies may be more easily accommodated than in others, depending on their degree of 'relative deprivation'; their view of what they can ultimately tolerate. It is interesting to note that the contemporary sociological notion of 'relative deprivation' (Runciman, 1972) can be found also in Marx's writings. In a series of commentaries in the *Neue Rheinische Zeitung,* starting in April 1849, he stated explicitly that our needs, being of a societal nature, are partly based on social comparisons and thus of a relative nature (Marx, *Wage Labour and Capital*).

With regard to the last three prerequisites mentioned in

our general statement, it would seem much easier to arrive at precise concepts and operationalizations. 'The strength of demands for societal change' relates both to the *numbers* demanding such change, and to the *social consciousness* of those involved, and their level of *organizational strength*. With regard to numbers it is necessary to determine whether or not to count the middle-strata of white-collar employees as belonging to the working class. More recent developments in advanced industrial societies have exposed this middle-stratum to some of the typical experiences of labour: mechanization, and threats of unemployment or dislocation. These developments would suggest that the size of the working class, in a sociological sense, may be relatively constant or even increasing in spite of the fact that the white-collar stratum has been growing at the expense of the numbers of blue-collar workers.

Another consideration with regard to the numbers of those making demands from below relates to the sociology of marginalization. In some countries — for instance in Latin America (Lindqvist, 1974: 213-14) — marginalization means a dislocation of unemployed or 'unemployables' into isolated and dispersed positions of weakness. Marginalization then would seem to go hand in hand with demobilization. Whether marginalization in advanced industrial societies is combined with demobilization, or whether the so-called 'sheltered employment' often provided through government support for this category of workers helps them to keep their discontent alive and shared in functioning collectivities, is a question which must be answered empirically before we can tell whether such marginalization reduces the numbers of those demanding or potentially demanding societal changes in such societies.

With regard to *type of social consciousness* I will say only this. Numbers alone provide no strength. But if you are numerous, and share a social consciousness conceiving society as unjust and opposed, and your own position in society as exposed to significant injustices, such consciousness can be transformed into an organized material force changing the course of history. Nor will I discuss 'the strength or weakness of established resistance to societal change' at greater length. Here I will only indicate some of the main dimensions of this concept. Obviously, numbers are less important here than such factors as *strategic power* and the appearance or reality of *indispensability* of such positions. In addition the factor of *social consciousness* is significant also at the establishment level. Crucial on this point is the degree of dissensus or consensus concerning modes of production and decision-making within these establishments. Here it would be particularly interesting to reformulate and empirically test some of the, perhaps, rather far-fetched ideas of James Burnham (1941) on the 'managerial revolution'. The current so-called STAMOCAP and STINCAP theories regarding relations between state and monopoly capital may also provide important points of departure for a critical elaboration of hypotheses regarding the strength of government and economic establishments within advanced capitalist societies.

Finally we have the question whether the *capabilities of implementing socialism* both at popular and administrative levels have been allowed to mature in the 'womb' of at least some advanced capitalist societies. An answer to this question requires both a reasonably precise and contemporary definition of socialism, and an empirical assessment of the relevant capabilities. Relevant capabilities would seem to be a seemingly rather unlikely combination of, for instance, a well-educated working-class and white-collar stratum, and central organizational strength of blue-collar and white-collar unions as well as labour parties, on the one hand, but also of decentralized workers' protests and wild-cat strikes which do not fragment and destroy centralized union power but put pressure on this power structure, and some experience among workers already within the capitalist system with regard to orderly participation in self-governing groups and responsible control of the working-environment and some aspects of management. It is quite possible to estimate empirically the existence of such 'un-likely' combinations. By way of conjecture, to be refuted or supported by such empirical studies, I venture to say that such combinations are least unlikely where reformist social democracy has had an enduring impact without, how-ever, having succeeded in changing the basic features of the capitalist system.

At the political and administrative or 'bureaucratic' levels the required capabilities for socialism are a proven willing-ness to enter into responsive dialogue with lower levels, and understanding as well as sympathy for the 'progressive' side in contradictions both within the superstructure and within the economic base. None of these capabilities, at whatever level, are terribly difficult to ascertain with fairly conven-tional methods of sociological research. Data are already available from previous studies and official statistics on some of these capabilities such as educational level and experience of democratic participation. Other capabilities must be investigated by new research.

EDITORIAL NOTE

We publish here a section relating more particularly to the scientific and technological revolution from a paper written by Professor Himmelstrand on the basis of the comments which he made at the end of the plenary session.

REFERENCES

Burnham, James (1941) *The Managerial Revolution* (New York).

Lindqvist, Sven (1974) *Jordens gryning. Jord och makt i Sydamerka,* del 2 (Bonniers, Stockholm).

Runciman, W.G. (1972) *Relative Deprivation and Social Justice* (Pelican, London).